Lecture Notes in Bioinformatics

T0237966

Subseries of Lecture Notes in Computer Science

Corrado Priami (Ed.)

Transactions on Computational Systems Biology IX

 Springer

Series Editors

Sorin Istrail, Brown University, Providence, RI, USA
Pavel Pevzner, University of California, San Diego, CA, USA
Michael Waterman, University of Southern California, Los Angeles, CA, USA

Editor-in-Chief

Corrado Priami
The Microsoft Research - University of Trento
Centre for Computational and Systems Biology
Piazza Manci, 17, 38050 Povo (TN), Italy
E-mail: priami@dit.unitn.it

Library of Congress Control Number: 2008937904

CR Subject Classification (1998): J.3, F.1.1, I.6, I.1

LNCS Sublibrary: SL 8 – Bioinformatics

ISSN   1861-2075
ISBN-10   3-540-88764-4 Springer Berlin Heidelberg New York
ISBN-13   978-3-540-88764-5 Springer Berlin Heidelberg New York

Springer is a part of Springer Science+Business Media

springer.com

© Springer-Verlag Berlin Heidelberg 2008

Typesetting: Camera-ready by author, data conversion by Scientific Publishing Services, Chennai, India
Printed on acid-free paper   SPIN: 12532667   06/3180   5 4 3 2 1 0

Preface

This issue of the journal reports regular papers. The first contribution is by Paras Chopra and Andreas Bender and discusses quantitative modelling aspects of the bgl operon for E.coli. The second contribution is by Rodrick Wallace and Deborah Wallace and deals with ecosystem transitions affecting phenotype expressions and selection mechanisms. The techniques used are statistical models. The third contribution is by Roberto Barbuti, Andrea Maggiolo-Schettini, Paolo Milazzo, Paolo Tiberi and Angelo Troina and presents the Stochastic Calculus of Looping Sequences (SCLS) suitable for the description of microbiological systems, such as cellular pathways, and their evolution. The last contribution is by Federica Ciocchetta and describes the use of biological transactions to make atomic sequences of interactions in the BlenX language.

May 2008   Corrado Priami

LNCS Transactions on Computational Systems Biology – Editorial Board

Table of Contents

A Quantitative *bgl* Operon Model for *E. coli* Requires BglF Conformational Change for Sugar Transport

Paras Chopra[1,2] and Andreas Bender[3]

[1] Molecular Reproduction, Development & Genetics, Indian Institute of Science,
Bangalore 560012, India
[2] Department of Biotechnology, Delhi College of Engineering, Delhi 110042, India
[3] Division of Medicinal Chemistry, Leiden / Amsterdam Center for Drug Research,
Gorlaeus Laboratory, Leiden University, Einsteinweg 55, 2300 Leiden,
The Netherlands
paras.chopra@bt.dce.edu
http://www.paraschopra.com

Abstract. The *bgl* operon is responsible for the metabolism of β-glucoside sugars such as salicin or arbutin in *E. coli*. Its regulatory system involves both positive and negative feedback mechanisms and it can be assumed to be more complex than that of the more closely studied *lac* and *trp* operons. We have developed a quantitative model for the regulation of the *bgl* operon which is subject to *in silico* experiments investigating its behavior under different hypothetical conditions. Upon administration of 5mM salicin as an inducer our model shows 80-fold induction, which compares well with the 60-fold induction measured experimentally. Under practical conditions 5-10mM inducer are employed, which is in line with the minimum inducer concentration of 1mM required by our model. The necessity of BglF conformational change for sugar transport has been hypothesized previously, and in line with those hypotheses our model shows only minor induction if conformational change is not allowed. Overall, this first quantitative model for the *bgl* operon gives reasonable predictions that are close to experimental results (where measured). It will be further refined as values of the parameters are determined experimentally. The model was developed in Systems Biology Markup Language (SBML) and it is available from the authors and from the Biomodels repository [www.ebi.ac.uk/biomodels].

Keywords: *bgl* operon, operon, transcriptional regulation, mathematical modeling, SBML model, biochemical modeling.

1 Introduction

The transcriptional control of gene expression is one of the major regulatory mechanisms by which both prokaryotic and eukaryotic organisms distinguish different phases of both the cell cycle [1] as well as during development [2], and

C. Priami (Ed.): Trans. on Comput. Syst. Biol. IX, LNBI 5121, pp. 1–22, 2008.
© Springer-Verlag Berlin Heidelberg 2008

one of the ways to adapt to the particular demands imposed by different environments. Nuclear hormone receptors are an example of receptors influencing gene transcription directly, and they have also been exploited as drug development targets, for example as the target of contraceptives. Modeling factors which influence gene transcription is a crucial step to understand living organisms. This is true on a fundamental level; but knowing how gene transcription is regulated in an organism could, for example in case of parasites, also lead to the discovery of novel drug targets which may be suitable points of modulation.

In this work, we will present, to the knowledge of the authors, the first computational model for *bgl* transcriptional regulation that combines experimental observations into a coherent way to give, on a small set of model perturbations, results which are close to experimental results which have been obtained until this stage.

We will in the following introduce the experimental evidence that is known today regarding the transcriptional regulation of the *bgl* operon. The model is discussed fully in the section "Computational Model Derivation of bgl Transcription", followed by a discussion of results obtained from the model.

An operon is a set of genes which are transcribed together to produce a single messenger RNA (mRNA) [3]. It is one of the most prominent (and at the same time simplest) strategies for genetic regulation in prokaryotes; having a single promoter, all the constituting genes in an operon are controlled simultaneously. Two of the best known and well studied examples of operons are the *lac* operon, which is involved in the metabolism of lactose, as well as the *trp* operon, involved in metabolism of trypthophane. Literature on simulating and modeling these operons abounds; for a general literature review on modeling and simulation strategies for genetic regulatory networks see a recent publication [4]. Specifically, a comprehensive mathematical modeling of the *lac* operon can be found in Yildirum *et al.* [3,4], while the *trp* operon has been simulated by *Santillan et al.* [5]. This paper extends the list of operons modeled by the first quantitative model for the *bgl* operon in *E. coli*.

The *bgl* operon is a cryptic operon [6], where by default the operon is in the silent state which disallows transcription. Only upon mutation the operon can become constitutively active, allowing for its transcription and regulatory control. The fact that no deletion of the gene occurred despite being a cryptic operon hints at the possibility that the *bgl* operon confers growth advantage under certain conditions [7,8]. This work is concerned with the active version of the operon.

Structurally, the *bgl* operon in *E. coli* consists of three protein-encoding genes [9] (see Figure 1): *bglF*, which is involved in the transport of β-glucoside sugars such as salicin and arbutin; *bglB*, which is responsible for hydrolyzing the transported sugar; and *bglG*, which acts as a positive regulator of the operon. Even though the *bgl* operon is inducible by β-glucoside sugars, the mechanism of regulation is different from other inducible operons such as the *lac* operon. From a regulatory point of view, the *bgl* operon encodes two proteins with opposing behavior: BglG acts to increase the expression of the operon, while BglF

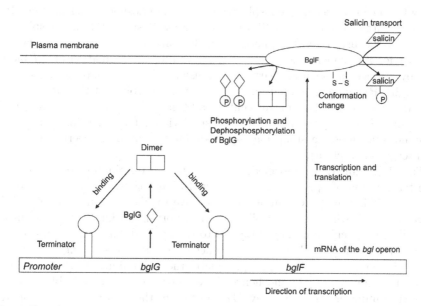

Fig. 1. Regulatory system of the *bgl* operon. (The *bglB* gene is omitted since it is only involved in sugar hydrolysis and not directly in regulation of *bgl*.) It is with our current knowledge likely to be more complex than that of the *lac* and *trp* operon due to the interplay between BglG, BglF and RNA terminators. See main text for a description of the regulatory mechanism of this operon.

interacts with BglG and renders it incapable of this function, thus inhibiting expression of the *bgl* gene. In contrast to the simple regulatory mechanism of *lac* and *trp* operons, this dual control mechanism of *bgl* expression possesses a comparatively complex structure.

In the work presented here, salicin was used as an inducer which was also employed in previous experimental settings [10]. Salicin, a β-glycoside sugar, does not have a direct effect on *bgl* regulation, but it gets phosphorylated as it is transported across the membrane. In turn, this phosphorylation of salicin results in a dephosphorylation of BglG monomers, which further leads to the formation of BglG dimers which act as anti-terminators of *bgl* transcription and, thereby, higher transcription levels.

The mechanism for *bgl* transcriptional regulation has been studied extensively in experiments [10,11,12,13]. The operon has two transcriptional terminators sandwiching the gene *bglG*, which is the first gene of the three to be transcribed (see Figure 1). The terminators cause a majority of RNA polymerases transcribing the operon to halt transcription, producing incomplete transcripts. Nevertheless, a small number of RNA polymerases go on to produce a full transcript which is translated to the proteins BglG, BglF and BglB. The regulatory system of the operon, considered in the model, is composed of only two components, BglG and BglF. BglB is only involved in sugar hydrolysis and has no regulatory function. See Figure 1 for an overview of the regulatory mechanism.

BglG, once produced, is phosphorylated by the phosphocarrier protein HPr, which is part of the bacterial Phosphotransferase System (PTS). Phosphorylation by HPr causes BglG to dimerise [14]. BglF also needs to be phosphorylated by HPr for its proper functioning [15]. Regarding the regulation of *bgl* induction by β-glucoside sugars, apart from transferring a phosphate group to BglG and BglF, HPr does not play any other significant role and thus it is not included in the model. This means that, at all times, the proteins BglG and BglF are assumed to be present in their phosphorylated form (for justifications of this and other model assumptions see "Computational Model Derivation of bgl Transcription" section).

The BglG dimer, once formed, can bind to the terminators that results in anti-termination of operon's transcription [16] and, thus, higher levels of transcription. This way, BglG can act as a positive regulator. BglF is a part of the PTS in *E. coli* which is responsible for transporting all kinds of sugars inside the cell. The specific function of BglF is to transport β-glucoside sugars such as salicin or arbutin [11]. However, in absence of any β-glucoside sugar in the extracellular environment BglF phosphorylates the BglG dimer [17,18], converting it into two monomers [19] and hence the dimer is no longer available to act as an anti-terminator [11]. This way, BglF decreases the expression of the operon in absence of β-glucoside sugars and, thus, it acts as a negative regulator.

While transporting a sugar molecule, BglF can transfer a phosphate group to it in two ways. Firstly (and most commonly), the phosphate transfer can occur through the PTS system. Secondly, BglF can dephosphorylate the already phosphorylated BglG and it can then transfer the available phosphate group to the sugar [11]. Thereby, when sugar is present, BglF allows the formation of BglG dimer. The dimer can, then, act as an anti-terminator and thus, the expression of the *bgl* operon is increased in presence of β-glucoside sugars. For a review on bgl's transcription anti-termination, see [20]. Also see [21] for a review on the structural basis of BglG's regulation.

As described above BglF can perform two opposite functions, both the phosphorylation and the dephosphorylation of BglG. It has been suggested that BglF acts as a molecular switch which can be stimulated by β-glucosides which causes it to change its role from phosphorylating BglG to dephosphorylating it [22]. The detailed molecular mechanism of the switching action is yet to be investigated. Preliminary research, however, reveals that the sugar transformation is concerted with the conformational change of BglF so that it can no longer phosphorylate the BglG dimer. This conformation change involves the disulphide bond formation between two cysteine residues at its active site [22]. Investigating the importance of allowing BglF conformational change is one of the central points of the current work, as outlined and analyzed later.

Apart from the regulation of the *bgl* operon directly by transcription antitermination, a distinct regulatory pathway has also been discovered, namely catabolite repression [23]. As described, HPr, which itself is part of the PTS system, phosphorylates BglG which causes the formation of BglG dimer. Since HPr is part of the PTS system, it can directly control the degree of BglG's phosphorylation depending upon types of sugars available to the bacterium. We do not consider catabolite repression in our model for two reasons. On the practical side, we needed to reduce complexity of the model to a reasonable degree. On the scientific side, excluding catabolite repression from the model represents a situation when only β-glucoside sugars are available in the medium, which is precisely the situation when the *bgl* operon is expressed and not repressed - the the situation we are attempting to model. On the other hand that means that our model is not an accurate model of the behavior of *bgl* operon under more common conditions when both glucose and other sugars are available (but when *bgl* is inactivated).

To the best of the authors' knowledge, the work presented here is the first comprehensive attempt to model the regulation of the *bgl* operon. While experimentally the operon has been investigated before (as outlined above), no attempt has been made yet to integrate the factors influencing *bgl* transcription into a single model.

For the interested reader, Kremling *et al.* recently presented [24] a mathematical model for the catabolite repression in *E. coli*. More generally, comprehensive modeling of carbohydrate uptake and metabolism has been performed by the same group [25].

2 Computational Model Derivation of *bgl* Transcription

2.1 Model Approximations

We have made several assumptions while designing the model. Firstly, the growth rate of the bacterium has been neglected so as to reduce the complexity of the model. Secondly, as exact quantitative data was not available, the concentration of all species except genes, RNA polymerase and ribosome was assumed to be zero at the start of the simulations, while the inducer, salicin, is only introduced once the species take a steady state concentration. Thirdly, some backward reactions such as unbinding of the RNA polymerase from an open complex were not considered because they are very slow in comparison to their respective forward reactions. For instance, the dissociation rate of BglG Dimer and Terminator 1 (GG.Ter1) complex is 10^6 times slower than the corresponding association rate [26]. Fourthly, in all experiments except one, the concentration of the extracellular β-glucoside sugar (salicin) was taken as constant over time. This is also the case in real-world experiments to a sufficient degree.

2.2 Model Equations

The model consists of a set of deterministic chemical reactions. The important reactions for the regulation of the operon include the interactions between RNA polymerase and terminators, BglF and BglG, BglF and salicin, and BglG dimer and terminators. The reactions are defined by their kinetic parameters. Except for few parameters where direct kinetic values were available we needed to estimate the values of most of the parameters. Table 1 lists the reactions and their corresponding kinetic parameters while Table 2 lists the constants chosen for the model with their experimental, where known, or the estimated values. The transcription of bgl by RNA polymerase follows a serial route. It means that the first terminator (Ter1) can only be transcribed once RNA polymerase has made a complex with the promoter region. Similarly, $bglG$ can only be transcribed once Ter1 has been transcribed. This is represented in the model as follows. $RNApol_{cell}$ is the free RNA polymerase available in the cell. It binds with the $bglR$ region, which is the promoter for the bgl operon, and the complex thus formed is represented by $RNApol_{bglR}$ in the model (Table 1, Reaction 1). $RNApol_{bglR}$ interacts with DNA_{Ter1} (DNA coding for the first terminator) to make Ter1 and $RNApol_{Ter1}$. ($RNApol_{Ter1}$ represents RNA Polymerase after transcribing Ter1.) $RNApol_{Ter1}$ is a precursor of $mRNA_G$ (mRNA coding for BglG; Reaction 2). This way, $mRNA_G$ can only be produced after the first terminator has been produced (Reaction 3). Both the incompletely transcribed transcript and fully transcribed transcript of the bgl operon contain a ribosome binding site for $bglG$. Hence, BglG is produced from both of the transcripts and, thus, in the model, $mRNA_G$ is produced corresponding to both of the transcripts. Only the fully transcribed transcript has a ribosome binding site for $bglF$, therefore $mRNA_F$ (mRNA coding for BglF) is produced corresponding only to the production of a complete bgl transcript. Hence in the simulation results, the actual bgl operon's expression level (concentration of the fully transcribed bgl transcript) is represented by the concentration of $mRNA_F$ (Reaction 4). The first terminator has already slowed the transcription in the previous step; the second terminator further slows down the process. Both, $mRNA_G$ and $mRNA_F$ are produced in this reaction because the full bgl operon's transcript codes for both BglG and BglF proteins $mRNA_G$ now binds to the ribosome, which gives the $mRNA_G$/ribosome complex (Reaction 5). This complex produces the translated protein BglG (plus unbound ribosome and $mRNA_G$) with the rate constant K_{tran_G}. $mRNA_F$ on the other hand also forms a complex with the ribosome (Reaction 6) and is translated with rate constant K_{tran_F} to give BglF (which is already phorphorylated), as well as again free ribosome and free $mRNA_F$. BglG is now able to dimerise with rate constant K_{dimer} (Reaction 7), while BglF is able to interact with the salicin inducer as follows (Reaction 8): Firstly, BglF (which is still phosphorylated) interacts with salicin to form a complex. This complex can either dissociate (second part of Reaction 8) or transport the sugar into the cell, with rate constant K_{salin}. If the latter takes place, BglF will form a disulphide bond which introduces conformational change (third part of Reaction 8). This disulphide bridge will be cleaved in cellular environments by thioredoxin and glutaredoxin, as Reaction 10 outlines. The BglG dimer on the other hand undergoes a two-step reversible reaction which is shown in Reaction 9. In the first step, an equilibrium exists between phosphorylated BglF and the BglG dimer as individual monomers, and the complex between BglF and the BglG dimer. In the second step, phosphorylated BglF is able to phosphorylate the BglG dimer, which leads to phosphorylated BglG monomers. On the other hand, the BglG dimer can also undergo a second reaction (Reaction 11), namely complex formation with the terminators, both Ter1 and Ter2. In both cases, this increases the transcription rate because now the terminators are unable to block the RNA polymerase. For all species, degradation is considered in the model with the reactions outlined in Reaction 12.

Table 1. Reactions constituting the model and their kinetic parameters

S. No.	Chemical reactions and their kinetic parameters
1.	$$RNApol_{cell} + bglR \xrightarrow{Kr\_cell} RNApol_{bglR}$$ $RNApol_{cell}$ is the RNA Polymerase in cell. $RNApol_{bglR}$ is the RNA Polymerase attached with the *bglR*.
	$$Kr\_cell = 6.41 \times 10^4 M^{-1} sec^{-1}; \ [RNApol_{cell}] = 2.6\mu M$$ [Operon] = 2.075nM; [bglR], [bglG], [bglF] and the [DNATer1]= [Operon]
2.	$$RNApol_{bglR} + DNA_{Ter1} \xrightarrow{Kter1} Ter1 + RNApol_{Ter1}$$ $$RNApol_{Ter1} \xrightarrow{Krnap\_diss} RNApol_{cell} + DNA_{Ter1}$$ DNA_{Ter1} is the DNA coding for first terminator (Ter1). The second reaction shows the dissociation of RNA Polymerase when the transcription halts due to blockage by the terminators. RNApol$_{Ter1}$ represents RNA Polymerase after transcribing Ter1.
	$$Kter1 = 5.87 \times 10^8 M^{-1} sec^{-1}$$ $$Krnap\_diss = 0.01 sec^{-1}$$
3.	$$RNApol_{Ter1} + bglG \xrightarrow{KbglG \times \frac{Kinb1}{Kinb1+[Ter1]}} mRNA_G + Ter2 + RNApol_G$$

Table 1. (*continued*)

The second factor $\frac{Kinb1}{Kinb1+[Ter1]}$ in reaction kinetics decreases the rate of reaction as the concentration of first terminator increases. Thus, it accounts for the activity of first terminator. Only $mRNA_G$ is produced in this reaction because the incomplete transcript codes only for the BglG.

$$KbglG = 2.245 \times 10^7 M^{-1} sec^{-1}; \; Kinb1 = 0.264nM$$

4.

$$RNApol_G + bglF \xrightarrow{KbglF \times \frac{Kinb2}{Kinb2+[Ter2]}} mRNA_F + mRNA_F + RNApol_{cell}$$

The first terminator has already slowed the transcription in previous step; the second terminator further slows down the process. Both, $mRNA_G$ and $mRNA_F$ are produced in this reaction because the full bgl operon's transcript codes for both BglG and BglF proteins.

$$KbglF = 9.83 \times 10^6 M^{-1} sec^{-1}; \; Kinb2 = 0.366nM$$

5.

$$mRNA_G + Ribosome \xrightarrow{Kg\_ribo} mRNA_G \cdot Ribo$$

$$mRNA_G \cdot Ribo \xrightarrow{Ktran\_G} G + Ribosome + mRNA_G$$

The first reaction represents the binding of Ribosome with mRNA. The second step represents the elongation step. Here, the protein BglG is represented by 'G'.

$$Kg\_ribo = 10^8 M^{-1} sec^{-1}; \; Ktran\_G = 0.048 sec^{-1}$$

6.

$$mRNA_F + Ribosome \xrightarrow{Kf\_ribo} mRNA_F \cdot Ribo$$

$$mRNA_F \cdot Ribo \xrightarrow{Ktran\_F} F^* + Ribosome + mRNA_F$$

These reactions represent the production of BglF. The enzyme BglF is represented by F^* to show that it is already phosphorylated by Hpr through the Phosphotransferase system (PTS) system.

Table 1. (*continued*)

	$Kf\_ribo = 10^8 M^{-1} sec^{-1}; \; Ktran\_F = 0.023 sec^{-1}$
7.	$G + G \xrightarrow{K_{dimer}} GG$
	This reaction represents the dimerization of BglG.
	$K_{dimer} = 3860 \times 10^6 M^{-1} sec^{-1}$
8.	$F^* + Salicin_{out} \xrightarrow{K_{salout}} Salicin_{out} \cdot F^*$
	$Salicin_{out} \cdot F^* \xrightarrow{K_{-salout}} F^* + Salicin_{out}$
	$Salicin_{out} \cdot F^* \xrightarrow{K_{salin}} Salicin_{in}^* + F_{S-S}^*$
	The above reactions represent the transport of β-glucoside sugar(s) such as salicin. Once the sugar has been transported, the conformation of BglF changes so that it may not phosphorylate the BglG dimer. This conformation change is represented by change of F^* to F_{S-S}^* which now has disulphide bonds between two of its cysteine residues [22].
	$K_{salout} = 1.04 \times 10^6 M^{-1} sec^{-1}; \; K_{-salout} = 1.55 sec^{-1};$ $K_{salin} = 19.2 sec^{-1}$
9.	$F^* + GG \xrightarrow{K_{phos\_G}} F^* \cdot GG$
	$F^* \cdot GG \xrightarrow{K_{-phos\_G}} F^* + GG$
	$F^* \cdot GG \xrightarrow{K_{phos\_G^*}} F^* + 2G^*$

Table 1. (*continued*)

$$F^* + 2G^* \xrightarrow{K^*_{-phos\_G}} F^* \cdot GG$$

The phosphorylation of the BglG dimer is shown by above reactions. It is a two step reaction in which both the steps are reversible.

$$K_{phos\_G} = 0.83 \times 10^6 M^{-1} sec^{-1}; \; K_{-phos\_G} = 1.55 sec^{-1}$$

$$K_{phos\_G^*} = 19.2 sec^{-1}; \; K^*_{-phos\_G^*} = 5.4 \times 10^6 M^{-1} sec^{-1}$$

| 10. | $$F^*_{S-S} \xrightarrow{\frac{Vmax \times [S]}{Km+[S]}} F^*$$ |

The disulfide bonds are continuously reduced in the cellular environment by two main pathways: thioredoxin and glutaredoxin [22,1]. We model the enzymatic reaction which reduces the disulphide bonds in the thioredoxin pathway [31].

$$Vmax = 0.331 \mu M^{-1} sec^{-1}; \; Km = 13.4 \mu M$$

| 11. | $$GG + Ter1 \xrightarrow{K_{anti-ter1}} GG \cdot Ter1$$ |

$$GG + Ter2 \xrightarrow{K_{anti-ter2}} GG \cdot Ter2$$

When the BglG dimer forms a complex with the terminators, it increases the transcription rate because now the terminators are unable to block the RNA polymerase. Mathematically, the formation of this complex decreases the concentration of Ter1 & Ter2. When [Ter1], for instance, is decreased, the magnitude of the factor $\frac{Kinb1}{Kinb1+[Ter1]}$ is increased, which results in net increase of the transcription rate.

$$K_{anti-ter1} = 10^6 M^{-1} sec^{-1}; \; K_{anti-ter2} = 10^6 M^{-1} sec^{-1}$$

Table 1. (*continued*)

12.	$mRNA_G, mRNA_F, Ter1, Ter2 \xrightarrow{K_{RNA\_degrad}}$
	$G, G^*, F^*, F^*_{S-S} \xrightarrow{K_{protein\_degrad}}$
	$GG \cdot Ter1, GG \cdot Ter2 \xrightarrow{K_{protein\_degrad}}$
	Representation of the degradation of various species over the time.
	$K_{RNA\_degrad} = 0.00235 sec^{-1}; K_{protein\_degrad} = 0.000385 sec^{-1}$

For a list of model equations see Table 1 and for the choice of constants with literature references see Table 2.

2.3   Computational Details

The model was developed in Systems Biology Markup Language (SBML) [27]. The SBML model file may be requested from the authors (P.C.). The software chosen for modeling was CellDesigner [28] while simulations were performed using Jarnac [29]. (For an overview of tools that can be used for kinetic modeling of biochemical networks see a recent review [30].)

Species not mentioned explicitly in Table 2 have initially zero concentrations.

Table 2. Model parameters and their corresponding values

S. No.	Parameter	Notes/Reference
1.	$Kr\_cell = 6.41 \times 10^4 M^{-1} sec^{-1}$	Used the value of transcription initiation rate for the trypthophan operon [5].
2.	$[RNApol_{cell}] = 2.6 \mu M$	Concentration of the RNA polymerase in vivo [32].

Table 2. (*continued*)

3.	$[Operon] = 2.075nM$	Calculated as follows. Assuming the volume of an *E. coli* cell to be 8×10^{-16} liters [5]. (Note that only one copy of the *bgl* operon is present in the wild type cell.) Now, the concentration of one copy (or molecule) is $\frac{1}{(6.023 \times 10^{2}3)(8 \times 10^{-16})} = 2.075nM$. Note: Here, the 'Operon' represents bglR, bglG, bglF and the DNATer1.
4.	$Kter1 = 5.87 \times 10^{8} M^{-1} sec^{-1}$	Calculated as follows. The length of DNA_{Ter1} is 32bp and the transcription rate in *E. coli* is 39 nt/sec [5]. Therefore, the time taken for the transcription of DNA_{Ter1} is 0.82 seconds.
5.	$KbglG = 2.245 \times 10^{7} M^{-1} sec^{-1}$	Calculated in the same way as done for calculating Kter1 above. In this case, the length of the *bglG* is 837bp.
6.	$Kinb1 = 0.264nM$	Calculated as follows. An 8.8 fold increase in operon expression is seen if first terminator is not present/non-functional [10]. So, taking $\frac{Kinb2}{Kinb2+[Ter2]} = \frac{1}{8.8}$ and setting [Ter1]=2.075nM, we get Kinb1=0.264nM.
7.	$KbglF = 9.83 \times 10^{6} M^{-1} sec^{-1}$	Calculated in the same way as done for calculating Kter1 above. In this case, the length of $bglF + DNA_{Ter2}$ is 1910 bp.
8.	$Kinb2 = 0.366nM$	Calculated in the same way as done for calculating Kter1 above. In this case, a 6.7 fold increase in operon expression is seen if second terminator is functionally inactive [10].
9.	$Kg\_ribo = 10^{8} M^{-1} sec^{-1}$	Taken from [33].

Table 2. (*continued*)

10.	$Ktran\_G = 0.048sec^{-1}$	Calculated as follows. The length of $mRNA_G$ is 278 amino acids (AAs) and the transcription rate in *E. coli* is 15 AAs/sec [34]. Therefore, total time taken for synthesis of G is 20.5 seconds. The total time includes the time spent (2 seconds) for translation initiation step also.
11.	$Kf\_ribo = 10^8 M^{-1} sec^{-1}$	Taken from [33].
12.	$Ktran\_F = 0.023sec^{-1}$	Calculated in the same way as done for calculating $Ktran\_G$ above. In this case, the length of $mRNA_F$ is of 625 AAs.
13.	$K_{dimer} = 3860 \times 10^6 M^{-1} sec^{-1}$	Taken from [35].
14.	$K_{salout} = 1.04 \times 10^6 M^{-1} sec^{-1}$; $K_{-salout} = 1.55sec^{-1}$; $K_{salin} = 19.2sec^{-1}$	Calculated [34,35,36]. The salicin transport rate in *E. coli* is 24% of the glucose transport rate, therefore kinetic parameters involved in salicin transport is taken to be 0.24 times the corresponding parameters in glucose transport.
15.	$K_{phos\_G} = 0.83 \times 10^6 M^{-1} sec^{-1}$	Taken from [18].
16.	$K_{-phos\_G} = 1.55sec^{-1}$; $K_{phos\_G*} = 19.2sec^{-1}$; $K_{-phos\_G*} = 5.4 \times 10^6 M^{-1} sec^{-1}$	Calculated using data from [36]. Treating the phosphorylation & dephosphorylation of GG analogous to that of glucose in PTS and then using the same kinetic values for GG as for glucose transport.
17.	$K_{anti-ter1} = 10^6 M^{-1} sec^{-1}$; $K_{anti-ter2} = 10^6 M^{-1} sec^{-1}$	Assumed values. Standard values of RNA binding proteins fall in this range.
18.	$Vmax = 0.331\mu M^{-1} sec^{-1}$; $Km = 13.4\mu M$	Taken from [36]. The values correspond to the enzymatic reaction in *Helicobacter pylori* which reduces the disulphide bonds in thioredoxin pathway
19.	$K_{RNA\_degrad} = 0.00235sec^{-1}$	Calculated. Assumed average half life of *bgl* operon's transcript as 294 seconds [37].
20.	$K_{protein\_degrad} = 0.000385sec^{-1}$	Calculated. The Standard half life of a protein in *E. coli* is 30 minutes.
21.	$Krnap\_diss = 0.01sec^{-1}$	Taken from [38]

3   Results and Discussion

3.1   Steady State Levels with and without Inducer

The model is simulated over a period of time and the concentration of species under observation (mRNA$_F$ in this case) is tracked. During the course of simulation, when the value becomes constant with respect to time, the value is said to be steady-state value of the species' concentration. At the start of the simulation, the concentrations of all species (except RNA polymerase, ribosome, and Genes) was taken to be zero. The results of this steady-state experiment are listed in Table 3, both for settings without inducer and at a concentration of the inducer salicin of 5mM. It can be seen that the bgl transcript concentration is increased 80-fold in the presence of salicin, while the concentration of phosphorylated BglF is decreased to the millionth part. The BglG concentration as well as the phosphorylated BglG concentration is virtually unchanged, while more dimer (concentration increased by a factor of 10^5) is present. Concentrations of the BglG dimer with the first and second terminator complex are increased between 10^3 and 10^5 fold.

3.2   Induction of the *bgl* Operon

Upon administration of 5mM salicin as an inducer, an 80 fold increase (Figure 2 and Table 3) in the bgl operon (mRNA$_F$) expression level could be observed. Experimentally, a 60 fold induction has been reported by Schnetz and Rak [10] at the same inducer concentration. Given the large number of estimated parameters of the model both numbers are roughly of the same order of magnitude.

Table 3. Steady state levels of different species when (a) no inducer is present, and (b) when 5mM inducer is present

Inducer Concentration	(a) No inducer	(b) 5mM
Species	Conc. (M)	Conc. (M)
bgl transcript (ribosome unbound) (mRNA$_F$)	3.90×10^{-10}	2.26×10^{-08}
BglF phosphorylated (F*)	1.38×10^{-05}	6.43×10^{-11}
BglF changed conformation (F*$_{s-s}$)	0	1.49×10^{-05}
BglG (G)	1.68×10^{-09}	1.59×10^{-09}
BglG Dimer (GG)	1.20×10^{-12}	1.86×10^{-07}
BglG phosphorylated (G*)	5.68×10^{-05}	4.77×10^{-05}
BglG Dimer & first terminator complex (GG.Ter1)	4.53×10^{-10}	8.77×10^{-07}
BglG Dimer & second terminator complex (GG.Ter2)	3.80×10^{-12}	3.26×10^{-07}

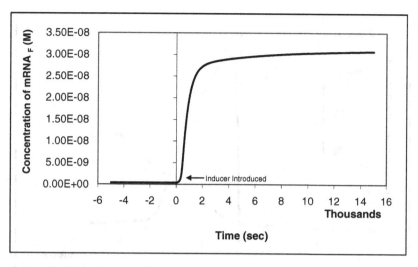

Fig. 2. An 80-fold induction of *bgl* expression is seen at an inducer (salicin) concentration of 5mM. The concentration of mRNA$_F$ (representative of the *bgl* full transcript) reached a plateau at about 310 nM, representing about 80 times the basal level of mRNA$_F$ without inducer. This compares to 60-fold induction seen experimentally at this inducer concentration [10].

3.3 The Dynamics of Induction When Inducer Is Present in Limiting Amounts

The biologically most appropriate regulatory behavior would boost the operon expression only as long as the inducer is available. The operon should not remain at high expression levels when the inducer is exhausted in the medium due to its consumption or otherwise. To check if the *bgl* operon is optimal in this regard, we performed this hypothetical experiment with our model. Unlike previous experiments where the inducer was throughout present at a constant concentration, the inducer in this experiment could be depleted (and ultimately get exhausted) as it was being transported across the membrane. The model predicts a rate of transport of the inducer (salicin) by BglF of $\sim 0.31 \mu M/s$. Therefore, to carry out this experiment for 20,000 seconds, $\sim 0.31 \mu M/s \times 20,000 = 6.2mM$ of inducer was introduced at time zero. It can be observed in Figure 3 that initially the expression level of the bgl operon rises considerably (~ 80 fold) and then when salicin gets exhausted, expression gradually falls to basal levels within the expected timeframe of 20,000 seconds.

3.4 Inducer Concentration *vs.* Induction Intensity

We next determined the minimum concentration of the inducer needed for any significant induction in the operon. This experiment involved the determination of the increase in the operon's expression levels at different inducer concentrations. The results are shown in Figure 4. It can be observed that there is no

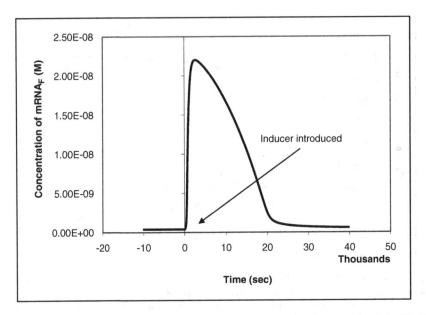

Fig. 3. Induction of *bgl* expression at limiting amounts of inducer (6.2 mM). The inducer in this experiment does not have a constant concentration. It is being depleted, hence limiting, as it is being transported by BglF. At t=0, 6.2 mM inducer are introduced, leading to rapid induction followed by relaxation to basal expression levels.

significant induction of the inducer up to concentration of around 1mM. This value of 1mM, hence, is the minimum concentration of the inducer to cause any significant induction. As most of the induction experiments are performed at 5-10 mM inducer concentration [10,11,12,13], the minimum inducer concentration given by our model is in agreement with experimental procedures.

3.5 Sensitivity of the Model to the Kinetic RNA-Binding Parameters Kter1 and Kter2

The parameters Kter1 and Kter2 were assigned the value for kinetic rates of standard RNA binding proteins $(10^6 M^{-1} sec^{-1})$ [26]. This was necessary since no kinetic data for BglG-terminator binding has yet been reported. To investigate the sensitivity of our model to these assumed kinetic parameters we now examined the sensitivity of the model to variations of Kter1 and Kter2 which were changed simultaneously to 10^{-2}, 10^{-1}, 10^1 and 10^2 fold the initially assumed value. We then observed the changes in the *bgl* transcript's concentration when (a) no inducer and (b) 5mM inducer were present. Results are shown in Figure 5. If no inducer was present, induction was independent of this parameter. At an inducer concentration of 5mM, induction did depend on the particular value of this constant, but to a much lesser than linear degree. For our model this means that the precise choice of Kter1 and Kter2 is less critical and that we can be confident to obtain satisfactory model behavior with the chosen parameters.

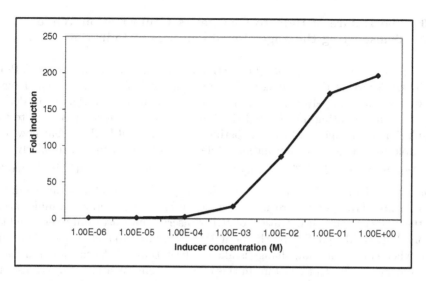

Fig. 4. Induction of *bgl* expression at different inducer concentrations. The results suggest that at least a concentration of 1mM is needed for significant induction. Since in practice experiments are usually carried out at concentrations between 5mM and 10mM, this is in agreement with our model.

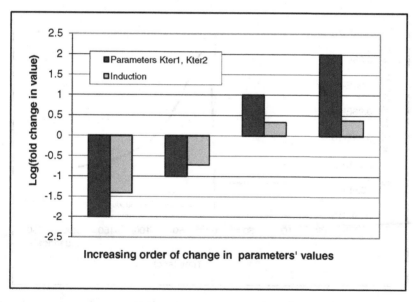

Fig. 5. Sensitivity of the model for the protein-RNA dissociation constants Kter1 and Kter2 at an inducer concentration of 5mM. An n-fold parameter change results in a much smaller change in *bgl* induction. Without inducer the model is completely insensitive to simultaneous changes of Kter1 and Kter2.

3.6   The Failure of BglF to Change Its Conformation While Transporting the Sugar Results in Loss of Induction

One of the regulatory steps of *bgl* induction is the conformational change in BglF while transporting the sugar [22]. This phenomenon of conformational change, however, is still largely a hypothesis and has not been investigated in detail. We were investigating how crucial the conformational change in BglF is to the regulation of the operon. In this experiment, we mutated BglF so that now it is not able to change its conformation while transporting sugar. Particularly, the reaction $Salicin_{out} \cdot F^* \xrightarrow{K_{salin}} Salicin_{in}^* + F_{S-S}^*$ (Table 1, Reaction 8) in the model was modified to $Salicin_{out} \cdot F^* \xrightarrow{K_{salin}} Salicin_{in}^* + F^*$. Results of this modified system are shown in Figure 6. It can be observed that upon administration of the inducer the concentration of bgl transcript increases only very slightly (from 3.9E-10 M to 3.95E-10 M; about 1%), compared to a change of about 80-fold when conformational change is allowed (see Table 2). Even though inducer concentrations are kept constant in this case, the concentration of bgl transcript falls to pre-inducer levels very quickly. Thus we can conclude that our model confirms the requirement of a conformational change of BglF proposed by Chen *et al.* [22].

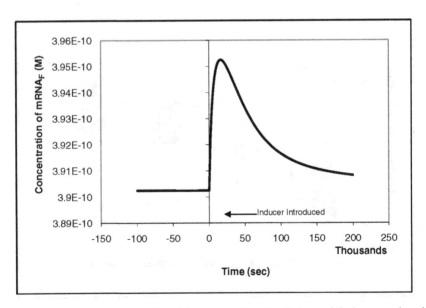

Fig. 6. Induction dynamics of the *bgl* operon when BglF is modified not to be able to undergo a conformational change while transporting sugar. Initially, upon inducer administration, a minimal (∼ 1%) induction can be observed. in the *bgl* transcript concentration, followed by relaxation to basal values. Unlike in Figure 3 the inducer concentration is constant throughout the simulation. This result supports that BglF conformational change is required to keep bgl expression and constantly high levels.

3.7 Mutations in *bglG* and *bglF* Make the Expression of bgl Operon Constitutive

It has been shown experimentally that loss-of-function mutations in either the *bglG* or *bglF* genes cause constitutive expression of the bgl operon [12,13]. Thus, both of the components of this regulatory system are required for normal regulation. If in our model system *bglF* is mutated, we observe a 200-fold change of *bgl* expression, while *bglG* mutations lead to basal expression levels. (Data are given in Table 4.) In both cases, expression is independent of the presence of inducer. This is in agreement with experimental observation, where constitutive expression results in expression independent of inducer concentrations. Functionally speaking, the *bglG* mutation results in the inability of this protein

Table 4. Change in operon's expression when *bglG* or *bglF* is mutated. The expression becomes constitutive when these genes are mutated. That is, it becomes independent of the presence of inducer.

With no inducer	
Mutation in	Fold change in expression
bglF	~200
bglG	1 (basal level)
With 5mM inducer	
Mutation in	Fold change in expression
bglF	~ 200
bglG	1 (basal level)

Table 5. Predictions made by the model and experimental status

Experiment No.	Prediction	Experimental Status
1.	No oscillatory expression of the operon seen in the model.	Unknown.
2.	80 fold induction at 5mM inducer.	60 fold induction at 5mM inducer [10].
3	The lower threshold for the inducer concentration for induction is ~1mM.	Most of the induction experiments are carried at 5-10 mM inducer [10,13].
4	The conformational change in BglF is an essential part of the regulation.	The role of conformation change of BglF has been suggested in [22].
5	Loss-of-function mutation in either BglF or BglG make the operon's expression constitutive.	Confirmed in [12,13].
7	Steady state concentration levels of various species.	Unknown.

to form a dimer, while a *bglF* mutation results in its inability to transport the inducer and phosphorylate the BglG dimer.

3.8   Predictions Made by the Model

Table 5 summarizes all the predictions made by the model, as discussed in the previous sections, along with their respective experimental status. It can be seen that for conditions where experimental data are present our model gives results which agree with experimental data. Further experiments performed in the future will be used to further refine this model.

4   Conclusions

The *bgl* operon is involved in the metabolism of β-glucoside sugars such as salicin or arbutin which also act as its inducers. In this work we present the first quantitative model for the regulation of the bgl operon. Upon administration of 5mM inducer our model shows 80-fold induction, which compares well with the 60-fold induction measured experimentally. In practice usually 5-10mM inducer are employed, which is in agreement with the minimum inducer concentration necessary in our model of 1mM. The necessity of BglF conformational change for sugar transport has been hypothesized previously, and our model shows only minor induction if this conformational change is not allowed. Overall, this first model for the bgl operon gives reasonable predictions that are close to experimental results (where measured). Given continuously available new experimental data, we will update the model in the future and ensure its consistency.

Acknowledgements. The Indian Academy of Sciences Summer Research Fellowship 2006 Program is thanked for funding the stay of Paras Chopra in Prof. S. Mahadevan's laboratory. Paras Chopra wishes to thank Prof. S. Mahadevan, V. Nanjundiah, and Ms. Ranjna Madan of Molecular Reproduction, Development & Genetics, Indian Institute of Science for their generous help and insightful suggestions.

References

1. Sanchez, I., Dynlacht, B.D.: Transcriptional control of the cell cycle. Curr. Opin. Cell. Biol. 8, 318–324 (1996)
2. Paul, J.: Transcriptional control during development. Biosci. Rep. 2, 63–76 (1982)
3. Jacob, F., Monod, J.: Genetic Regulatory Mechanisms in Synthesis of Proteins. J. Mol. Biol. 3, 318 (1961)
4. De Jong, H.: Modeling and simulation of genetic regulatory systems: A literature review. J. Comput. Biol. 9, 67–103 (2002)
5. Santillan, M., Mackey, M.C.: Dynamic regulation of the tryptophan operon: A modeling study and comparison with experimental data. Proc. Natl. Acad. Sci. U. S. A. 98, 1364–1369 (2001)

6. Hall, B.G., Yokoyama, S., Calhoun, D.H.: Role of Cryptic Genes in Microbial Evolution. Mol. Biol. Evol. 1, 109–124 (1983)

7. Madan, R., Kolter, R., Mahadevan, S.: Mutations that activate the silent bgl operon of Escherichia coli confer a growth advantage in stationary phase. J. Bacteriol. 187, 7912–7917 (2005)

8. Mukerji, M., Mahadevan, S.: Cryptic genes: evolutionary puzzles. Journal of Genetics 76, 147–159 (1997)

9. Henkin, T.M.: Control of transcription termination in prokaryotes. Annu. Rev. Genet. 30, 35–57 (1996)

10. Schnetz, K., Rak, B.: Regulation of the Bgl Operon of Escherichia-Coli by Transcriptional Antitermination. Embo. J. 7, 3271–3277 (1988)

11. Amster-Choder, O., Houman, F., Wright, A.: Protein-Phosphorylation Regulates Transcription of the Beta-Glucoside Utilization Operon in Escherichia-Coli. Cell 58, 847–855 (1989)

12. Mahadevan, S., Wright, A.: A Bacterial Gene Involved in Transcription Antitermination - Regulation at a Rho-Independent Terminator in the Bgl Operon of Escherichia-Coli. Cell 50, 485–494 (1987)

13. Mahadevan, S., Reynolds, A.E., Wright, A.: Positive and Negative Regulation of the Bgl Operon in Escherichia-Coli. J. Bacteriol. 169, 2570–2578 (1987)

14. Ben-Zeev, E., Fux, L., Amster-Choder, O., Eisenstein, M.: Experimental and computational characterization of the dimerization of the PTS-regulation domains of BglG from Escherichia coli. J. Mol. Biol. 347, 693–706 (2005)

15. Postma, P.W., Lengeler, J.W.: Phosphoenolpyruvate - Carbohydrate Phosphotransferase System of Bacteria. Microbiol. Rev. 49, 232–269 (1985)

16. Gorke, B., Rak, B.: Efficient transcriptional antitermination from the Escherichia coli cytoplasmic membrane. J. Mol. Biol. 308, 131–145 (2001)

17. Gorke, B.: Regulation of the Escherichia coli antiterminator protein BglG by phosphorylation at multiple sites and evidence for transfer of phosphoryl groups between monomers. J. Biol. Chem. 278, 46219–46229 (2003)

18. Lopian, L., Nussbaum-Shochat, A., O'Day-Kerstein, K., Wright, A., Amster-Choder, O.: The BglF sensor recruits the BglG transcription regulator to the membrane and releases it on stimulation. Proc. Natl. Acad. Sci. U. S. A. 100, 7099–7104 (2003)

19. Fux, L., Nussbaum-Shochat, A., Amster-Choder, O.: A fraction of the BglG transcriptional antiterminator from Escherichia coli exists as a compact monomer. J. Biol. Chem. 278, 50978–50984 (2003)

20. Amster-Choder, O.: The bgl sensory system: a transmembrane signaling pathway controlling transcriptional antitermination. Curr. Opin. Microbiol. 8, 127–134 (2005)

21. van Tilbeurgh, H., Le Coq, D., Declerck, N.: Crystal structure of an activated form of the PTS regulation domain from the LicT transcriptional antiterminator. Embo. J. 20, 3789–3799 (2001)

22. Chen, Q., Nussbaum-Shochat, A., Amster-Choder, O.: A novel sugar-stimulated covalent switch in a sugar sensor. J. Biol. Chem. 276, 44751–44756 (2001)

23. Gorke, B., Rak, B.: Catabolite control of Escherichia coli regulatory protein BglG activity by antagonistically acting phosphorylations. Embo. J. 18, 3370–3379 (1999)

24. Kremling, A., Bettenbrock, K., Laube, B., Jahreis, K., Lengeler, J.W., Gilles, E.D.: The organization of metabolic reaction networks. III. Application for diauxic growth on glucose and lactose. Metab. Eng. 3, 362–379 (2001)

25. Kremling, A., Fischer, S., Sauter, T., Bettenbrock, K., Gilles, E.D.: Time hierarchies in the Escherichia coli carbohydrate uptake and metabolism. Biosystems 73, 57–71 (2004)

26. Harada, Y., Funatsu, T., Murakami, K., Nonoyama, Y., Ishihama, A., Yanagida, T.: Single-molecule imaging of RNA polymerase-DNA interactions in real time. Biophys. J. 76, 709–715 (1999)

27. Hucka, M., Finney, A., Sauro, H.M., Bolouri, H., Doyle, J.C., Kitano, H., Arkin, A.P., Bornstein, B.J., Bray, D., Cornish-Bowden, A., Cuellar, A.A., Dronov, S., Gilles, E.D., Ginkel, M., Gor, V., Goryanin II, Hedley, W.J., Hodgman, T.C., Hofmeyr, J.H., Hunter, P.J., Juty, N.S., Kasberger, J.L., Kremling, A., Kummer, U., Le Novere, N., Loew, L.M., Lucio, D., Mendes, P., Minch, E., Mjolsness, E.D., Nakayama, Y., Nelson, M.R., Nielsen, P.F., Sakurada, T., Schaff, J.C., Shapiro, B.E., Shimizu, T.S., Spence, H.D., Stelling, J., Takahashi, K., Tomita, M., Wagner, J., Wang, J.: The systems biology markup language (SBML): a medium for representation and exchange of biochemical network models. Bioinformatics 19, 524–531 (2003)

28. http://celldesigner.org/

29. http://sbw.kgi.edu/software/jarnac.htm

30. Alves, R., Antunes, F., Salvador, A.: Tools for kinetic modeling of biochemical networks. Nat. Biotechnol. 24, 667–672 (2006)

31. Baker, L.M.S., Raudonikiene, A., Hoffman, P.S., Poole, L.B.: Essential thioredoxin-dependent peroxiredoxin system from Helicobacter pylori: Genetic and kinetic characterization. J. Bacteriol. 183, 1961–1973 (2001)

32. Bremer, H., Dennis, P.P.: Escherichia Coli and Salmonella Typhimurium. Cellular and Molecular Biology (1987)

33. Kierzek, A.M., Zaim, J., Zielenkiewicz, P.: The effect of transcription and translation initiation frequencies on the stochastic fluctuations in prokaryotic gene expression. J. Biol. Chem. 276, 8165–8172 (2001)

34. Hall, B.G.: Predicting evolutionary potential. I. Predicting the evolution of a lactose-PTS system in Escherichia coli. Mol. Biol. Evol. 18, 1389–1400 (2001)

35. Fux, L., Nussbaum-Shochat, A., Amster-Choder, O.: Interactions between the PTS regulation domains of the BglG transcriptional antiterminator from Escherichia coli. J. Biol. Chem. 278, 46203–46209 (2003)

36. Rohwer, J.M., Meadow, N.D., Roseman, S., Westerhoff, H.V., Postman, P.W.: Understanding glucose transport by the bacterial phosphoenolpyruvate: glycose phosphotransferase system on the basis of kinetic measurements in vitro. J. Biol. Chem. 275, 34909–34921 (2000)

37. Gulati, A., Mahadevan, S.: The Escherichia coli antiterminator protein BglG stabilizes the 5' region of the bgl mRNA. J. Biosci. 26, 193–203 (2001)

38. McKnight, S.L., Yamamoto, K.R.: Transcriptional regulation. Cold Spring Harbor Laboratory Press, Plainview (1992)

Punctuated Equilibrium in Statistical Models of Generalized Coevolutionary Resilience: How Sudden Ecosystem Transitions Can Entrain Both Phenotype Expression and Darwinian Selection

Rodrick Wallace[1,*] and Deborah Wallace[2]

[1] The New York State Psychiatric Institute
wallace@pi.cpmc.columbia.edu
[2] Consumers Union
rdwall@ix.netcom.com

Abstract. We argue that mesoscale ecosystem resilience shifts akin to sudden phase transitions in physical systems can entrain similarly punctuated events of gene expression on more rapid time scales, and, in part through such means, slower changes induced by selection pressure, triggering punctuated equilibrium Darwinian evolutionary transitions on geologic time scales. The approach reduces ecosystem, gene expression, and Darwinian genetic dynamics to a least common denominator of information sources interacting by crosstalk at markedly differing rates. Pettini's 'topological hypothesis', via a homology between information source uncertainty and free energy density, generates a regression-like class of statistical models of sudden coevolutionary phase transition based on the Rate Distortion and Shannon-McMillan Theorems of information theory which links all three levels. A mathematical treatment of Holling's extended keystone hypothesis regarding the particular role of mesoscale phenomena in entraining both slower and faster dynamical structures produces the result. A main theme is the necessity of a cognitive paradigm for gene expression, mirroring I. Cohen's cognitive approach to immune function. Invocation of the necessary conditions imposed by the asymptotic limit theorems of communication theory enables us to penetrate one layer more deeply before needing to impose an empirically-derived phenomenological system of 'Onsager relation' recursive coevolutionary stochastic differential equations. Extending the development to second order via a large deviations argument permits modeling the influence of human cultural structures on ecosystems as 'farming'.

1   Introduction

Early in the twentieth century, evolutionary biologists debated whether species change occurred gradually or as a result of massive catastrophes. At that time,

---

* Address correspondence to: R. Wallace, 549 W. 123 St., Suite 16F, New York, NY, 10027. Telephone (212) 865-4766. Affiliations are for identification only.

C. Priami (Ed.): Trans. on Comput. Syst. Biol. IX, LNBI 5121, pp. 23–85, 2008.

the gradualists prevailed, and the catastrophists were marginalized. Speciation was viewed as a gradual process of incremental changes in response to incremental environmental challenges. However, Eldredge and Gould [28, 42], after study of the fossil record, concluded that speciation occurred suddenly. Species appeared in the fossil record, remained in the fossil record largely unchanged, and then disappeared. There was little or no evidence of gradual incremental changes that led to speciation. Eldredge and Gould called the process that they saw 'punctuated equilibrium', a term that referred to the sudden changes (punctuations) and the quiet interims (equilibria), a combination of gradualism and catastrophism. Eldredge and Gould published their initial findings in the early 1970's.

In the same period C. S. Holling developed the ecosystem equivalent of punctuated equilibrium, namely ecosystem resilience theory [35, 44, 47, 81]. Resilience theory views each ecosystem as normally in a quasi-equilibrium state. As the ecosystem is subjected to various impacts, it shows no obvious changes in structure or function but the relationships between the species become tighter as the perturbations erode the more delicate peripheral relationships. Finally, either a more intense impact occurs or the aggregated impacts over time shatter so many loose relationships that those remaining become brittle and shatter. The ecosystem then shifts relatively suddenly into a different dynamic domain, a different quasi-equilibrium with markedly different structure and function. Examples of domain change include natural ones such as change of forest into prairie after drought and major forest fires in areas marginal for forests and unnatural ones such as eutrophication of waterbodies from agricultural runoff and discharge of urban wastewater. A great body of empirical work supports this perspective [44].

Ecosystems provide the niches for species. If ecosystems are suddenly transformed into different configurations, then species are confronted with sudden changes in selection pressures. It seems likely that Holling's theory provides something of an explanation for Eldredge and Gould's reading of the fossil record. Besides the fossil record, the climatological and geological records also show major changes in temperature, atmospheric composition, and geological processes such as volcanoes, earthquakes, and movements of tectonic plates. These, of course, form the macroscale of ecosystems. Local topography, geology, hydrology, and microclimate lead to ecological niches. Organisms by their activities modify their own niches and the niches of other organisms [55, 64]. These localized processes form the microscale of ecosystems. Landscape processes such as wildfires which spread and affect large numbers of niches form the mesoscale [48].

Niches within ecosystems select for the fittest phenotypes for them. Not all genes of an organism are expressed. Thus, the genetic variability within the population of a particular niche may be far greater than the relatively uniform phenotype presented to the examining ecologist. If a characteristic may potentially be influenced by multiple genes, the niche may select for a phenotype consonant with the expression of only a single, or a very few, genes. The species in the fossil record reflect only phenotypes, not the full range of genetic

variability. Ecosystem domain shift would lead to selection for different phenotypes. Those individuals with the genes that can express these newly 'preferred'
phenotypes will supplant the old phenotypes in the new ecosystem configuration, producing apparent speciation. As time hardens the new ecoconfiguration,
the genetic composition of the 'new species' will indeed shift toward the old
alleles and new mutations expressing the new phenotypes most efficiently, and
true speciation occurs. The book *Animal Traditions* [8] describes in detail how
behavioral phenotypes can become encoded in the genome, as does Ancel's analysis of the Baldwin effect [4] in which the ability to learn becomes convoluted
with gene expression and selection.

Here we will analyze the interaction of these phenomena using a principled
approach which reduces ecosystem dynamics, gene expression, and Darwinian genetic selection to a least common denominator of interacting information sources
constrained by the asymptotic limit theorems of information theory. This is not
an entirely new perspective. Priami [73], for example, finds that the interaction
between biological entities can be represented as an exchange of information between programs. Earlier, Jimenez-Montano [49] and Waddington [82] had suggested that language may become a paradigm for a theory of general biology,
but a language in which basic sentences are programs, not simple statements.
Our particular contribution is to hew very closely indeed to the basic mathematical structure of the asymptotic limit theorems of information theory and the
associated generalizations afforded by the large deviations program of applied
probability.

We will begin by first placing ecosystem dynamics, gene expression, and Darwinian gene selection on a similar footing as expressions of different information
sources. We will then examine the interaction between information sources, using the homology between information source uncertainty and the free energy
density of a physical system to import phase transistion methods from statistical
physics via Pettini's [68, 69] topological hypothesis. An analog to the Onsager
relations of nonequilibrium statistical mechanics permits study of these systems
far from phase transisiton, leading to a coevolutionary paradigm which extends
much contemporary analysis focused on genes alone. The reexpression of Ancel's
work on the Baldwin effect [4] in terms of a 'tuning theorem' variant of the Shannon coding theorem produces the essential result that mesoscale ecosystem shifts
will be particularly powerful in entraining gene expression and gene selection.

2   Ecosystems as Information Sources

2.1   Coarse-Graining a Simple Model

We begin with a simplistic picture of an elementary predator/prey ecosystem
which, nonetheless, provides a useful pedagogical starting point. Let X represent
the appropriately scaled number of predators, Y the scaled number of prey, t
the time, and ω a parameter defining their interaction. The model assumes that
the ecologically dominant relation is an interaction between predator and prey,
so that

$$dX/dt = \omega Y \tag{1}$$
$$dY/dt = -\omega X$$

Thus the predator populations grows proportionately to the prey population, and the prey declines proportionately to the predator population.

After differentiating the first and using the second equation, we obtain the differential equation

$$d^2X/dt^2 + \omega^2 X = 0 \tag{2}$$

having the solution

$$X(t) = sin(\omega t); Y(t) = cos(\omega t).$$

with

$$X(t)^2 + Y(t)^2 = sin^2(\omega t) + cos^2(\omega t) \equiv 1.$$

Thus in the two dimensional phase space defined by $X(t)$ and $Y(t)$, the system traces out an endless, circular trajectory in time, representing the out-of-phase sinusoidal oscillations of the predator and prey populations.

Divide the $X - Y$ phase space into two components – the simplest coarse graining – calling the halfplane to the left of the vertical Y-axis A and that to the right B. This system, over units of the period $1/(2\pi\omega)$, traces out a stream of A's and B's having a very precise grammar and syntax, i.e.

$$ABABABAB...$$

Many other such statements might be conceivable, e.g.

$$AAAAA..., BBBBB..., AAABAAAB..., ABAABAAAB...,$$

and so on, but, of the obviously infinite number of possibilities, only one is actually observed, is 'grammatical', i.e. $ABABABAB....$

More complex dynamical system models, incorporating diffusional drift around deterministic solutions, or even very elaborate systems of complicated stochastic differential equations, having various domains of attraction, i.e. different sets of grammars, can be described by analogous symbolic dynamics (e.g., [11], Ch. 3).

2.2 Ecosystems and Information

Rather than taking symbolic dynamics as a simplification of more exact analytic or stochastic approaches, it proves useful, as it were, to throw out the Cheshire cat, but keep the cat's smile, generalizing symbolic dynamics to a more comprehensive information dynamics: Ecosystems may not have identifiable sets of stochastic dynamic equations like noisy, nonlinear clocks, but, under appropriate coarse-graining, they may still have recognizable sets of grammar and syntax over the long-term.

Examples abound. The turn-of-the seasons in a temperate climate, for many natural communities, looks remarkably the same year after year: the ice melts, the migrating birds return, the trees bud, the grass grows, plants and animals reproduce, high summer arrives, the foliage turns, the birds leave, frost, snow, the rivers freeze, and so on.

Suppose it is indeed possible to empirically characterize an ecosystem at a given time t by observations of both habitat parameters such as temperature and rainfall, and numbers of various plant and animal species.

Traditionally, one can then calculate a cross-sectional species diversity index at time t using an information or entropy metric of the form

$$H = -\sum_{j=1}^{M}(n_j/N)\log[(n_j/N)],$$

$$N \equiv \sum_{j=1}^{M} n_j$$

where n_j is the number of observed individuals of species j and N is the total number of individuals of all species observed (e.g., [32, 70, 74]).

This is not the approach taken here. Quite the contrary, in fact. Suppose it is possible to coarse grain the ecosystem at time t according to some appropriate partition of the phase space in which each division A_j represent a particular range of numbers of each possible species in the ecosystem, along with associated parameters such as temperature, rainfall, and the like. What is of particular interest to our development is not cross sectional structure, but rather longitudinal paths, i.e. ecosystem statements of the form $x(n) = A_0, A_1, ..., A_n$ defined in terms of some natural time unit of the system, i.e. n corresponds to an again appropriate characteristic time unit T, so that $t = T, 2T, ..., nT$.

To reiterate, unlike the traditional use of information theory in ecology, our interest is in the *serial correlations along paths*, and not at all in the cross-sectional entropy calculated for of a single element of a path.

Let $N(n)$ be the number of possible paths of length n which are consistent with the underlying grammar and syntax of the appropriately coarsegrained ecosystem, e.g. spring leads to summer, autumn, winter, back to spring, etc. but never something of the form spring to autumn to summer to winter in a temperate ecosystem.

The fundamental assumptions are that – for this chosen coarse-graining – $N(n)$, the number of possible grammatical paths, is much smaller than the total number of paths possible, and that, in the limit of (relatively) large n,

$$H = \lim_{n\to\infty} \frac{\log[N(n)]}{n} \tag{3}$$

both exists and is independent of path.

This is a critical foundation to, and limitation on, the modeling strategy and its range of strict applicability, but is, in a sense, fairly general since it is *independent of the details of the serial correlations along a path.*

Again, these conditions are the essence of the parallel with parametric statistics. Systems for which the assumptions are not true will require special nonparametric approaches. We are inclined to believe, however, that, as for parametric statistical inference, the methodology will prove robust in that many systems will sufficiently fulfill the essential criteria.

This being said, not all possible ecosystem coarse-grainings are likely to work, and different such divisions, even when appropriate, might well lead to different descriptive quasi-languages for the ecosystem of interest. The example of Markov models is relevant. The essential Markov assumption is that the probability of a transition from one state at time T to another at time $T + \Delta T$ depends only on the state at T, and not at all on the history by which that state was reached. If changes within the interval of length ΔT are plastic, or path dependent, then attempts to model the system as a Markov process *within* the natural interval ΔT will fail, even though the model works quite well for phenomena separated by natural intervals.

Thus empirical identification of relevant coarse-grainings for which this body of theory will work is clearly not trivial, and may, in fact, constitute the hard scientific core of the matter.

This is not, however, a new difficulty in ecosystem theory. Holling [48], for example, explores the linkage of ecosystems across scales, finding that mesoscale structures – what might correspond to the neighborhood in a human community – are ecological keystones in space, time, and population, which drive process and pattern at both smaller and larger scales and levels of organization. This will, in fact, be a core argument of our development.

Levin [54] argues that there is no single correct scale of observation: the insights from any investigation are contingent on the choice of scales. Pattern is neither a property of the system alone nor of the observer, but of an interaction between them. Pattern exists at all levels and at all scales, and recognition of this multiplicity of scales is fundamental to describing and understanding ecosystems. In his view there can be no 'correct' level of aggregation: we must recognize explicitly the multiplicity of scales within ecosystems, and develop a perspective that looks across scales and that builds on a multiplicity of models rather than seeking the single 'correct' one.

Given an appropriately chosen coarse-graining, whose selection in many cases will be the difficult and central trick of scientific art, suppose it possible to define joint and conditional probabilities for different ecosystem paths, having the form $P(A_0, A_1, ..., A_n), P(A_n|A_0, ..., A_{n-1})$, such that appropriate joint and conditional Shannon uncertainties can be defined on them. For paths of length two these would be of the form

$$H(X_1, X_2) \equiv -\sum_j \sum_k P(A_j, A_k) \log[P(A_j, A_k)] \qquad (4)$$

$$H(X_1|X_2) \equiv -\sum_j \sum_k P(A_j, A_k) \log[P(A_j|A_k)],$$

where the X_j represent the stochastic processes generating the respective paths of interest.

The essential content of the Shannon-McMillan Theorem is that, for a large class of systems characterized as information sources, a kind of law-of-large numbers exists in the limit of very long paths, so that

$$H[X] = \lim_{n \to \infty} \frac{\log[N(n)]}{n} = \tag{5}$$
$$\lim_{n \to \infty} H(X_n | X_0, ..., X_{n-1}) =$$
$$\lim_{n \to \infty} \frac{H(X_0, X_1, ..., X_n)}{n + 1}.$$

Taking the definitions of Shannon uncertainties as above, and arguing backwards from the latter two equations (e.g., [51]), it is indeed possible to recover the first, and divide the set of all possible temporal paths of our ecosystem into two subsets, one very small, containing the grammatically correct, and hence highly probable paths, which we will call 'meaningful', and a much larger set of vanishingly low probability.

Basic material on information theory can be found in any number of texts, e.g., [5, 21, 51].

The next task is to show how the cognitive processes which so distinguish much individual and collective animal activity, as well as many basic physiological processes, can be fitted into a similar context, i.e. characterized as information sources.

3   Cognition as an Information Source

Atlan and Cohen [6] argue that the essence of cognition is comparison of a perceived external signal with an internal, learned picture of the world, and then, upon that comparison, the choice of one response from a much larger repertoire of possible responses. Such reduction in uncertainty inherently carries information, and, following the approach of [84, 89], it is possible to make a very general model of this process as an information source.

Cognitive pattern recognition-and-selected response, as conceived here, proceeds by convoluting an incoming external 'sensory' signal with an internal 'ongoing activity' – the learned picture of the world – and, at some point, triggering an appropriate action based on a decision that the pattern of sensory activity requires a response. It is not necessary to specify how the pattern recognition system is trained, and hence possible to adopt a weak model, regardless of learning paradigm, which can itself be more formally described by the Rate Distortion Theorem. Fulfilling Atlan and Cohen's (1998) criterion of meaning-from-response, we define a language's contextual meaning entirely in terms of system output.

The model, an extension of that presented in [89], is as follows.

A pattern of 'sensory' input, say an ordered sequence $y_0, y_1, ...$, is mixed in a systematic (but unspecified) algorithmic manner with internal 'ongoing' activity, the sequence $w_0, w_1, ...,$ to create a path of composite signals $x = a_0, a_1, ..., a_n, ...,$ where $a_j = f(y_j, w_j)$ for some function f. This path is then fed into a highly nonlinear, but otherwise similarly unspecified, decision oscillator which generates an output $h(x)$ that is an element of one of two (presumably) disjoint sets B_0 and B_1. We take

$$B_0 \equiv b_0, ..., b_k, \tag{6}$$
$$B_1 \equiv b_{k+1}, ..., b_m.$$

Thus we permit a graded response, supposing that if

$$h(x) \in B_0 \tag{7}$$

the pattern is not recognized, and if

$$h(x) \in B_1 \tag{8}$$

the pattern is recognized and some action $b_j, k + 1 \leq j \leq m$ takes place.

The principal focus of interest is those composite paths x which trigger pattern recognition-and-response. That is, given a fixed initial state a_0, such that $h(a_0) \in B_0$, we examine all possible subsequent paths x beginning with a_0 and leading to the event $h(x) \in B_1$. Thus $h(a_0, ..., a_j) \in B_0$ for all $0 \leq j < m$, but $h(a_0, ..., a_m) \in B_1$.

For each positive integer n let $N(n)$ be the number of grammatical and syntactic high probability paths of length n which begin with some particular a_0 having $h(a_0) \in B_0$ and lead to the condition $h(x) \in B_1$. We shall call such paths meaningful and assume $N(n)$ to be considerably less than the number of all possible paths of length n – pattern recognition-and-response is comparatively rare. We – again – assume that the longitudinal finite limit $H \equiv \lim_{n \to \infty} \log[N(n)]/n$ both exists and is independent of the path x. We will – not surprisingly – call such a cognitive process *ergodic*.

Note that disjoint partition of state space may be possible according to sets of states which can be connected by meaningful paths from a particular base point, leading to a natural coset algebra of the system, a groupoid. This is a matter of some mathematical importance pursued in [39, 84, 89].

It is thus possible to define an ergodic information source $\mathbf{X}$ associated with stochastic variates X_j having joint and conditional probabilities $P(a_0, ..., a_n)$ and $P(a_n | a_0, ..., a_{n-1})$ such that appropriate joint and conditional Shannon uncertainties may be defined which satisfy the relations of equation (5) above.

This information source is taken as *dual* to the ergodic cognitive process.

We reiterate that the Shannon-McMillan Theorem and its variants provide 'laws of large numbers' which permit definition of the Shannon uncertainties in terms of cross-sectional sums of the form $H = -\sum P_k \log[P_k]$, where the P_k constitute a probability distribution.

It is important to recognize that different quasi-languages will be defined by different divisions of the total universe of possible responses into various pairs of sets B_0 and B_1. Like the use of different distortion measures in the Rate

Distortion Theorem (e.g., [21]), however, it seems obvious that the underlying dynamics will all be qualitatively similar.

Nonetheless, dividing the full set of possible responses into the sets B_0 and B_1 may itself require higher order cognitive decisions by another module or modules, suggesting the necessity of choice within a more or less broad set of possible quasi-languages. This would directly reflect the need to shift gears according to the different challenges faced by the organism or social group. A critical problem then becomes the choice of a normal zero-mode language among a very large set of possible languages representing the excited states accessible to the system. This is a fundamental matter which mirrors, for isolated cognitive systems, the resilience arguments applicable to more conventional ecosystems, i.e. the possibility of more than one zero state to a cognitive system. Identification of an excited state as the zero mode becomes, then, a kind of generalized autoimmune disorder which can be triggered by linkage with external ecological information sources of structured psychosocial stress, a matter we explore at length elsewhere [84].

In sum, meaningful paths – creating an inherent grammar and syntax – have been defined entirely in terms of system response, as Atlan and Cohen [6] propose.

This formalism can easily be applied to the stochastic neuron in a neural network, as done in [92].

Ultimately it becomes necessary to parametize the information source uncertainty of the dual information source to a cognitive pattern recognition-and-response with respect to one or more variates, writing, e.g., $H[\mathbf{K}]$, where $\mathbf{K} \equiv (K_1, ..., K_s)$ represents a vector in a parameter space. Let the vector $\mathbf{K}$ follow some path in time, i.e. trace out a generalized line or surface $\mathbf{K}(t)$. We assume that the probabilities defining H, for the most part, closely track changes in $\mathbf{K}(t)$, so that along a particular piece of a path in parameter space the information source remains as close to stationary and ergodic as is needed for the mathematics to work. Between pieces we will, below, impose phase transition characterized by a renormalization symmetry, in the sense of [95]. See the Mathematical Appendix for further details.

Such an information source can be termed adiabatically piecewise stationary ergodic (APSE). To reiterate, the ergodic nature of the information sources is a generalization of the law of large numbers and implies that the long-time averages we will need to calculate can, in fact, be closely approximated by averages across the probability spaces of those sources. This is no small matter.

The reader may have noticed parallels with Dretske's speculations on the the role of the asymptotic limit theorems of information theory in constraining high level mental function [26, 84, 92].

Wallace [92] and Wallace and Fullilove [84] describe in some detail how, for larger animals, immune function, tumor control, the hypothalamic-pituitary-adrenal (HPA) axis (the flight-or-fight system), emotion, conscious thought, and embedding group (and sometimes cultural) structures are all cognitive in this simple sense. In general these cognitive phenomena will occur at far faster rates than embedding ecosystem changes.

It is worth a more detailed recounting of the arguments for characterizing a number of physiological subsystems as cognitive in the sense of this section.

3.1 Immune Cognition

Atlan and Cohen [6] have proposed an information-theoretic cognitive model of immune function and process, a paradigm incorporating cognitive pattern recognition-and-response behaviors analogous to those of the central nervous system. This work follows in a very long tradition of speculation on the cognitive properties of the immune system (e.g., [43, 71, 80]).

From the Atlan/Cohen perspective, the meaning of an antigen can be reduced to the type of response the antigen generates. That is, the meaning of an antigen is functionally defined by the response of the immune system. The meaning of an antigen to the system is discernible in the type of immune response produced, not merely whether or not the antigen is perceived by the receptor repertoire. Because the meaning is defined by the type of response there is indeed a response repertoire and not only a receptor repertoire.

To account for immune interpretation Cohen [17, 18] has reformulated the cognitive paradigm for the immune system. The immune system can respond to a given antigen in various ways, it has 'options.' Thus the particular response we observe is the outcome of internal processes of weighing and integrating information about the antigen. In contrast to Burnet's view of the immune response as a simple reflex, it is seen to exercise cognition by the interpolation of a level of information processing between the antigen stimulus and the immune response. A cognitive immune system organizes the information borne by the antigen stimulus within a given context and creates a format suitable for internal processing; the antigen and its context are transcribed internally into the 'chemical language' of the immune system.

The cognitive paradigm suggests a language metaphor to describe immune communication by a string of chemical signals. This metaphor is apt because the human and immune languages can be seen to manifest several similarities such as syntax and abstraction. Syntax, for example, enhances both linguistic and immune meaning.

Although individual words and even letters can have their own meanings, an unconnected subject or an unconnected predicate will tend to mean less than does the sentence generated by their connection.

The immune system creates a 'language' by linking two ontogenetically different classes of molecules in a syntactical fashion. One class of molecules are the T and B cell receptors for antigens. These molecules are not inherited, but are somatically generated in each individual. The other class of molecules responsible for internal information processing is encoded in the individual's germline.

Meaning, the chosen type of immune response, is the outcome of the concrete connection between the antigen subject and the germline predicate signals.

The transcription of the antigens into processed peptides embedded in a context of germline ancillary signals constitutes the functional 'language' of the immune system. Despite the logic of clonal selection, the immune system does not respond to antigens as they are, but to abstractions of antigens-in-context.

Cohen [19] provides a more recent perspective, focusing on inflammatory processes as maintenance in which the immune decision-making process uses strategies similar to those observed in the nervous system.

3.2   Tumor Control

We argue that the next larger cognitive submodule after the immune system must be a tumor control mechanism which may include immune surveillance, but clearly transcends it. Nunney [63] has explored cancer occurrence as a function of animal size, suggesting that in larger animals, whose lifespan grows as about the 4/10 power of their cell count, prevention of cancer in rapidly proliferating tissues becomes more difficult in proportion to size. Cancer control requires the development of additional mechanisms and systems to address tumorigenesis as body size increases – a synergistic effect of cell number and organism longevity. Nunney ([63], p. 497) concludes that this pattern may represent a real barrier to the evolution of large, long-lived animals and predicts that those that do evolve have recruited additional controls over those of smaller animals to prevent cancer.

Different tissues may have evolved markedly different tumor control strategies. All of these, however, are likely to be energetically expensive, permeated with different complex signaling strategies, and subject to a multiplicity of reactions to signals, including those related to psychosocial stress. Forlenza and Baum [36] explore the effects of stress on the full spectrum of tumor control in higher animals, ranging from DNA damage and control, to apoptosis, immune surveillance, and mutation rate. Elsewhere [88] we argue that this elaborate tumor control strategy, particularly in large animals, must be at least as cognitive as the immune system itself, which is one of its components. That is, some comparison must be made with an internal picture of a 'healthy' cell, and a choice made as to response: none, attempt DNA repair, trigger programmed cell death, engage in full-blown immune attack. This is, from the Atlan/Cohen perspective, the essence of cognition.

3.3   A Cognitive Paradigm for Gene Expression

While modes of genetic inheritance are assumed well understood since the Grand Evolutionary Synthesis of the early 20th Century, the mechanisms of gene activation, regulation, and expression remain largely hidden. A random reading of the literature illuminates a stark and increasingly mysterious landscape.

Liu and Ringner [58] find gene expression signatures consisting of tens to hundreds of genes determine different biological states and conclude that it is crucial to systematically analyze gene expression signatures in the context of signaling pathways.

Soyer et al. [79] find that, although massive network structures are associated with the biological signal transduction which allows a cell or organism to sense its environment and react accordingly, the experimental work needed to gather enough quantitative information to develop accurate mathematical models is highly labor intensive, so that the modeling of specific networks may be of limited use in developing a broad understand of the general properties of biological signaling networks.

One possible mathematical characterization of these difficulties is found in Sayyed-Ahmad et al. [77], who explore the basic conundrum in terms of a dynamic model. In their view the state of a cell is specified by a set of variables Ψ for which we know the governing equations and a set T which is at the frontier of our understanding (i.e., for which we do not know the governing equations). The challenge is that the dynamics of Ψ is given by a cell model, e.g.,

$$d\Psi/dt = G(\Psi, T(t), \Lambda) \tag{9}$$

in which the rate G depends not only on many rate and equilibrium constants Λ, but also on the time-dependent frontier variables $T(t)$. The descriptive variables, Ψ, can only be determined as a function of the unknown time courses $T(t)$. Thus the model cannot be simulated.

Liao et al [57] find that use of statistical methods on biological networks, such as principal component analysis, ignore the underlying network structures and provide decompositions based on a priori statistical constraints on the computed component signals. The resulting decomposition, in their view, provides a phenomenological model for the observed data and does not necessarily contain physically or biologically meaningful signals.

Baker and Stock [9], however, pose the questions in a more general manner, using an information metaphor in which understanding of signal transduction systems has focused on mechanisms that allow crosstalk between different information processing modalities. They particularly ask what are the decision making mechanisms by which a bacterium controls the activities of its genes and proteins to adapt to changing environmental conditions? That is, how is information converted into knowledge, and how is knowledge sorted, evaluated and combined to guide action, morphogenesis and growth?

O'Nuallain [67] provides an important perspective on this approach. In his view the categorical failure to solve the general problem of natural language processing by computer is prognostic of the future of gene expression work. After what seemed like a promising start, in his view, the field was stalled by an inability to handle, or even define coherently, 'contextual' factors. Currently, he continues, the field is gradually being taken over by Bayesian 'methods' that simply look for the statistical incidence of co-occurrence of lexical items in the source (analogous to gene) and target (analogous to protein) languages. Contextual factors in the case of gene expression include the bioenergetic status of the cell, a status that can be assessed properly only with painstaking work; yet it determines what genes are being turned on and off at any particular moment.

It seems clear that 18th Century dynamical models using 19th Century differential equation generalizations of equation (9) have little to offer in addressing fundamental questions of gene activation and regulation. More sophisticated work must clearly move in the direction of an Atlan/Cohen cognitive paradigm for gene expression, characterizing the processes, and their embedding contexts, in terms of nested sets of interacting dual information sources, whose behavior is constrained by the necessary conditions imposed by the asymptotic limit theorems of communications theory.

That is, properly coarse-grained and nested biochemical networks will have an observed grammar and syntax, and, limited by powerful probability limit theorems, such description can enable construction of robust and meaningful statistical models of gene expression which may then be used for real scientific inference.

In sum, generalizing symbolic dynamics to a more inclusive, and less restrictive, cognitive paradigm for gene expression in terms of the model of equations (5) - (7), while invoking the inherent complexities of topological groupoids described in [39, 84], seems likely to provide badly needed illumination for this dark and confusing realm.

Not uncharacteristically, I. Cohen and colleagues (e.g., [20]) have, in fact, already proposed something much in this direction, using a 'reactive system' paradigm for gene expression taken from computer models. Reactive systems, in their view, call our attention to their emergent properties. An emergent property of a system is a behavior of the system, taken as a whole, that is not expressed by any one of the lower-scale components that comprise it. Although Cohen and Harel [20] then attempt to develop a complicated computer modeling strategy to address such reactive systems, Cohen [19] describes the essential differences between them and conventional computer architecture in some detail. There is no external operator or programmer, no programs, algorithms or software distinct from the system's hardware, no central processing unit, no operating system, no formal mathematical logic, no termination criteria, since the system never stops, no verification procedures, and so on.

Zhu et al. [97], by contrast, take an explicit kinetic chemical reaction approach to gene expression involving delayed stochastic differential equations. They begin by coarse-graining multi-step biochemical processes with single-step delayed reactions. Thus their coarse-graining involves not only collapsing biochemical steps, but collapsing as well the inevitable associated serial correlations into a small number of 'time delays'. The key feature of their model is that the complex multiple-step biochemical processes, such as transcription, translation, and even the whole gene expression, are simplified to single-step time delayed reactions.

While there are sufficiently many gene expression mechanisms so that some of them, at least, will yield to this method, we are interested in those which are more complex, and indeed broadly cognitive, subject to emergent patterns which cannot be modeled simply as bifurcations of stochastically-perturbed mechanistic models.

Indeed, rather than pursuing the computer models that [20] and [97] invoke, here we will attempt to extend our statistical and dynamic analytic treatment of the cognitive paradigm to a structure incorporating gene expression in a broadly coevolutionary manner. As Richard Hamming so famously put it, "The purpose of computing is insight, not numbers", and analytic models offer transparency as well as insight. We will, however, recover a phenomenological formalism as a kind of generalized Onsager model, but at a later, and far more global, stage of structure. That is, invocation of the necessary conditions imposed by the limit theorems of communication theory enables us to penetrate one layer deeper before it becomes necessary to call for an empirically-determined phenomenological system of Onsager relation stochastic differential equations.

4 Darwinian Genetic Inheritance as an Information Source

Adami et al. [2] make a case for reinterpreting the Darwinian transmission of genetic heritage in terms of a formal information process. They assert that genomic complexity can be identified with the amount of information a sequence stores about its environment: genetic complexity can be defined in a consistent information-theoretic manner. In their view, information cannot exist in a vacuum and must have an instantiation. For biological systems the instantiation of information is DNA. To some extent it is the blueprint of an organism and thus information about its own structure. More specifically, it is a blueprint of how to build an organism that can best survive in its native environment, and pass on that information to its progeny. They assert that an organism's DNA thus is not only a 'book' about the organism, but also a book about the environment it lives in, including the species it co-evolves with. They identify the complexity of geonomes by the amount of information they encode about the world in which they have evolved.

Ofria et al. [65] continue in the same direction and argue that genomic complexity can be defined rigorously within standard information theory as the information the genome of an organism contains about its environment. From the point of view of information theory, it is convenient to view Darwinian evolution on the molecular level as a collection of information transmission channels, subject to a number of constraints. In these channels, they state, the organism's genomes code for the information (a message) to be transmitted from progenitor to offspring, and are subject to noise due to an imperfect replication process. Information theory is concerned with analysing the properties of such channels, how much information can be transmitted and how the rate of perfect information transmission of such a channel can be maximized.

Adami and Cerf [1] argue, using simple models of genetic structure, that the information content, or complexity, of a genomic string by itself (without referring to an environment) is a meaningless concept and a change in environment (catastrophic or otherwise) generally leads to a pathological reduction in complexity.

The transmission of genetic information is thus a contextual matter which involves operation of an information source which, according to this development, must interact with embedding (ecosystem) structures.

Such interaction is, as we show next, often highly punctuated.

5   Interacting Information Sources: Punctuated Crosstalk

Suppose that a cognitive, or Darwinian, information process of interest can be represented by a sequence of states in time, the path $x \equiv x_0, x_1, \dots$. Similarly, we assume an embedding ecosystem with which that process interacts can also be represented by a path $y \equiv y_0, y_1, \dots$. These paths are both very highly structured and, within themselves, are serially correlated and can, in fact, be represented by information sources $\mathbf{X}$ and $\mathbf{Y}$. We assume the process of interest and the embedding ecosystem interact, so that these sequences of states are not independent, but are jointly serially correlated. We can, then, define a path of sequential pairs as $z \equiv (x_0, y_0), (x_1, y_1), \dots$.

The essential content of the Joint Asymptotic Equipartition Theorem (JAEPT) version of the Shannon-McMillan Theorem is that the set of joint paths z can be partitioned into a relatively small set of high probability which is termed *jointly typical*, and a much larger set of vanishingly small probability. Further, according to the JAEPT, the *splitting criterion* between high and low probability sets of pairs is the mutual information

$$I(X, Y) = H(X) - H(X|Y) = H(X) + H(Y) - H(X, Y) \qquad (10)$$

where $H(X), H(Y), H(X|Y)$ and $H(X, Y)$ are, respectively, the Shannon uncertainties of X and Y, their conditional uncertainty, and their joint uncertainty. Again, see [5, 21] for mathematical details. As stated above, the Shannon-McMillan Theorem and its variants permit expression of the various uncertainties in terms of cross sectional sums of terms of the form $-P_k \log[P_k]$ where the P_k are appropriate direct or conditional probabilities. Similar approaches to neural process have been recently adopted by Dimitrov and Miller [25].

The high probability pairs of paths are, in this formulation, all equiprobable, and if $N(n)$ is the number of jointly typical pairs of length n, then, according to the Shannon-McMillan Theorem and its 'joint' variants,

$$I(X, Y) = \lim_{n \to \infty} \frac{\log[N(n)]}{n}. \qquad (11)$$

Generalizing the earlier language-on-a-network models of [85, 86], suppose there is a coupling parameter P representing the degree of linkage between the cognitive human subsystem of interest and the structured quasi-language of the embedding ecosystem, and set $K = 1/P$, following the development of those earlier studies. Then we have

$$I[K] = \lim_{n \to \infty} \frac{\log[N(K, n)]}{n}.$$

The essential homology between information theory and statistical mechanics lies in the similarity of this expression with the infinite volume limit of the free energy density. If $Z(K)$ is the statistical mechanics partition function derived from the system's Hamiltonian, then the free energy density is determined by the relation

$$F[K] = \lim_{V \to \infty} \frac{\log[Z(K)]}{V}. \qquad (12)$$

F is the free energy density, V the system volume and $K = 1/T$, where T is the system temperature.

Various authors argue at some length (e.g., [34, 76, 84, 92]) that this is indeed a systematic mathematical homology which, as described in the Appendix, permits importation of renormalization methods into information theory. Imposition of invariance under renormalization on the mutual information splitting criterion $I(X, Y)$ implies the existence of phase transitions analogous to learning plateaus or punctuated evolutionary equilibria in the relations between cognitive mechanism and the embedding ecosystem. An extensive mathematical treatment of these ideas is presented elsewhere (e.g., [68, 69, 84, 89, 91]) and in the Mathematical Appendix. A detailed example will be given in a subsequent section. Much of the uniqueness of the system under study will be expressed in the renormalization relations associated with that punctuation.

Elaborate developments are possible. From a the more limited perspective of the Rate Distortion Theorem, a selective corollary of the Shannon-McMillan Theorem, we can view the onset of a punctuated interaction between the cognitive process and embedding ecosystem as the literal writing of distorted image of those systems upon each other, Lewontin's [56] interpenetration:

Suppose that two (adiabatically, piecewise stationary, ergodic) information sources $\mathbf{Y}$ and $\mathbf{B}$ begin to interact, to talk to each other, i.e. to influence each other in some way so that it is possible, for example, to look at the output of $\mathbf{B}$ – strings b – and infer something about the behavior of $\mathbf{Y}$ from it – strings y. We suppose it possible to define a retranslation from the B-language into the Y-language through a deterministic code book, and call $\hat{\mathbf{Y}}$ the translated information source, as mirrored by $\mathbf{B}$.

Define some distortion measure comparing paths y to paths $\hat{y}$, $d(y, \hat{y})$ [21]. We invoke the Rate Distortion Theorem's mutual information $I(Y, \hat{Y})$, which is the splitting criterion between high and low probability pairs of paths. Impose, now, a parametization by an inverse coupling strength K, and a renormalization symmetry representing the global structure of the system coupling.

Extending the analyses, triplets of sequences, Y_1, Y_2, Z, for which one in particular, here Z, is the 'embedding context' affecting the other two, can also be divided by a splitting criterion into two sets, having high and low probabilities respectively. The probability of a particular triplet of sequences is then determined by the conditional probabilities

$$P(Y_1 = y^1, Y_2 = y^2, Z = z) = \Pi_{j=1}^{n} p(y_j^1 | z_j) p(y_j^2 | z_j) p(z_j). \tag{13}$$

That is, Y_1 and Y_2 are, in some measure, driven by their interaction with Z.

For large n the number of triplet sequences in the high probability set will be determined by the relation ([21], p. 387)

$$N(n) \propto \exp[n I(Y_1; Y_2 | Z)], \tag{14}$$

where splitting criterion is given by

$$I(Y_1; Y_2|Z) \equiv$$
$$H(Z) + H(Y_1|Z) + H(Y_2|Z) - H(Y_1, Y_2, Z).$$

It is then possible to examine mixed cognitive/adaptive phase transitions analogous to learning plateaus [91] in the splitting criterion $I(Y_1, Y_2|Z)$. We reiterate that these results are almost exactly parallel to the Eldredge/Gould model of evolutionary punctuated equilibrium [28, 29, 42].

The model is easily extended to any number of interacting information sources, $Y_1, Y_2, ..., Y_s$ conditional on an external context Z in terms of a splitting criterion defined by

$$I(Y_1, ..., Y_s|Z) = H(Z) + \sum_{j=1}^{s} H(Y_j|Z) - H(Y_1, ..., Y_s, Z), \qquad (15)$$

where the conditional Shannon uncertainties $H(Y_j|Z)$ are determined by the appropriate direct and conditional probabilities.

If we assume interacting information sources can be partitioned into three different sets, perhaps fast, X_i, medium, Y_j and slow Z_k relative transmission rates, then mathematical induction on this equation produces a complicated expression of the form

$$I(X_1, ..., X_i|Y_1, ..., Y_j|Z_1, ..., Z_k). \qquad (16)$$

In general, then, it seems fruitful to characterize the mutual interpenetration of cognitive biopsychosocial and non-cognitive ecosystem and genetic structures within the context a single, unifying, formal perspective summarized by a 'larger' information source, more precisely, invoking a mutual information between cognitive, genetic, and ecosystem information sources.

6 Dynamic Manifolds

A fundamental homology between the information source uncertainty dual to a cognitive process and the free energy density of a physical system arises, in part, from the formal similarity between their definitions in the asymptotic limit. Information source uncertainty can be defined as in equation (4). This is, as noted, quite analogous to the free energy density of a physical system, equation (12).

Feynman [34] provides a series of physical examples, based on Bennett's work, where this homology is, in fact, an identity, at least for very simple systems. Bennett argues, in terms of idealized irreducibly elementary computing machines, that the information contained in a message can be viewed as the work saved by not needing to recompute what has been transmitted.

Feynman explores in some detail Bennett's ideal microscopic machine designed to extract useful work from a transmitted message. The essential argument is that computing, in any form, takes work. Thus the more complicated a cognitive process, measured by its information source uncertainty, the greater its energy consumption, and the ability to provide energy is limited. Inattentional blindness, Wallace [84] argues, emerges as a thermodynamic limit on processing capacity in a topologically-fixed global workspace, i.e. one which has been strongly configured about a particular task.

Understanding the time dynamics of cognitive systems away from the kind of phase transition critical points described above requires a phenomenology similar to the Onsager relations of nonequilibrium thermodynamics. This will lead to a more general phase transition theory involving large-scale topological changes in the sense of Morse theory, summarized in the Mathematical Appendix.

If the dual source uncertainty of a cognitive process is parametized by some vector of quantities $\mathbf{K} \equiv (K_1, ..., K_m)$, then, in analogy with nonequilibrium thermodynamics, gradients in the K_j of the *disorder*, defined as

$$S \equiv H(\mathbf{K}) - \sum_{j=1}^{m} K_j \partial H / \partial K_j \qquad (17)$$

become of central interest.

Equation (17) is similar to the definition of entropy in terms of the free energy density of a physical system, as suggested by the homology between free energy density and information source uncertainty described above.

Pursuing the homology further, the generalized Onsager relations defining temporal dynamics become

$$dK_j/dt = \sum_i L_{j,i} \partial S / \partial K_i, \qquad (18)$$

where the $L_{j,i}$ are, in first order, constants reflecting the nature of the underlying cognitive phenomena. The L-matrix is to be viewed empirically, in the same spirit as the slope and intercept of a regression model, and may have structure far different than familiar from more simple chemical or physical processes. The $\partial S / \partial K$ are analogous to thermodynamic forces in a chemical system, and may be subject to override by external physiological driving mechanisms.

An essential contrast with simple physical systems driven by (say) entropy maximization is that cognitive systems make decisions about resource allocation, to the extent resources are available. That is, resource availability is a context for cognitive function, in the sense of Baars, not a determinant.

Equations (17) and (18) can be derived in a simple parameter-free covariant manner which relies on the underlying topology of the information source space implicit to the development. Cognitive, genetic, and ecosystem phenomena are, according to our development, to be associated with particular information sources, and we are interested in the local properties of the system near a particular reference state. We impose a topology on the system, so that, near a particular 'language' A, dual to an underlying cognitive process, there is (in some sense) an open set U of closely similar languages $\hat{A}$, such that $A, \hat{A} \subset U$. Note that it may be necessary to coarse-grain the system's responses to define these information sources. The problem is to proceed in such a way as to preserve the underlying essential topology, while eliminating 'high frequency noise'. The formal tools for this can be found, e.g., in Chapter 8 of [14].

Since the information sources dual to the cognitive processes are similar, for all pairs of languages $A, \hat{A}$ in U, it is possible to:

[1] Create an embedding alphabet which includes all symbols allowed to both of them.

[2] Define an information-theoretic distortion measure in that extended, joint alphabet between any high probability (i.e. grammatical and syntactical) paths in A and $\hat{A}$, which we write as $d(Ax, \hat{A}x)$ [21]. Note that these languages do not interact, in this approximation.

[3] Define a metric on U, for example,

$$\mathcal{M}(A, \hat{A}) = |\lim \frac{\int_{A,\hat{A}} d(Ax, \hat{A}x)}{\int_{A,A} d(Ax, A\hat{x})} - 1|, \tag{19}$$

using an appropriate integration limit argument over the high probability paths. Note that the integration in the denominator is over different paths within A itself, while in the numerator it is between different paths in A and $\hat{A}$.

Consideration suggests $\mathcal{M}$ is a formal metric, having

$$\mathcal{M}(A, B) \geq 0, \mathcal{M}(A, A) = 0, \mathcal{M}(A, B) = \mathcal{M}(B, A),$$

$$\mathcal{M}(A, C) \leq \mathcal{M}(A, B) + \mathcal{M}(B, C).$$

Other approaches to metric construction on U seem possible.

Structures weaker than a conventional metric would be of more general utility, but the mathematical complications are formidable [39].

Note that these conditions can be used to define equivalence classes of *languages*, where previously, in cognitive process, we could define equivalence classes of *states* which could be linked by high probability, grammatical and syntactical, paths to some base point. This led to the characterization of different information sources. Here we construct an entity, formally a topological manifold, *which is an equivalence class of information sources*. This is, provided $\mathcal{M}$ is a conventional metric, a classic differentiable manifold. We shall be interested in topological states within such manifolds, and in the possibilities of transition between manifolds.

Since H and $\mathcal{M}$ are both scalars, a 'covariant' derivative can be defined directly as

$$dH/d\mathcal{M} = \lim_{\hat{A} \to A} \frac{H(A) - H(\hat{A})}{\mathcal{M}(A, \hat{A})}, \tag{20}$$

where $H(A)$ is the source uncertainty of language A.

Suppose the system to be set in some reference configuration A_0.

To obtain the unperturbed dynamics of that state, impose a Legendre transform using this derivative, defining another scalar

$$S \equiv H - \mathcal{M}dH/d\mathcal{M}. \tag{21}$$

The simplest possible Onsager relation – here seen as an empirical, fitted, equation like a regression model – in this case becomes

$$d\mathcal{M}/dt = LdS/d\mathcal{M}, \tag{22}$$

where t is the time and $dS/d\mathcal{M}$ represents an analog to the thermodynamic force in a chemical system. This is seen as acting on the reference state A_0. For

$$dS/d\mathcal{M}|_{A_0} = 0, \tag{23}$$
$$d^2 S/d\mathcal{M}^2|_{A_0} > 0$$

the system is quasistable, a Black hole, if you will, and externally imposed forcing mechanisms will be needed to effect a transition to a different state. We shall explore this circumstance below in terms of topological considerations analogous to the concept of ecosystem resilience.

Conversely, changing the direction of the second condition, so that

$$dS^2/d\mathcal{M}^2|_{A_0} < 0,$$

leads to a repulsive peak, a White hole, representing a possibly unattainable realm of states.

Explicit parametization of $\mathcal{M}$ introduces standard – and quite considerable – notational complications (e.g., [7, 14]): Imposing a metric for different cognitive dual languages parametized by $\mathbf{K}$ leads to Riemannian, or even Finsler, geometries, including the usual geodesics. See the Mathematical Appendix for details.

The dynamics, as we have presented them so far, have been noiseless, while neural systems, from which we are abducting theory, are well known to be very noisy, and indeed may be subject to mechanisms of stochastic resonance. Equation (22) might be rewritten as

$$d\mathcal{M}/dt = LdS/d\mathcal{M} + \sigma W(t)$$

where σ is a constant and $W(t)$ represents white noise. Again, S is seen as a function of the parameter $\mathcal{M}$. This leads directly to a family of classic stochastic differential equations having the form

$$d\mathcal{M}_t = L(t, dS/d\mathcal{M})dt + \sigma(t, dS/d\mathcal{M})dB_t, \tag{24}$$

where L and σ are appropriately regular functions of t and $\mathcal{M}$, and dB_t represents the noise structure.

In the sense of Emery [30], this leads into deep realms of stochastic differential geometry and related topics. The obvious inference is that noise, which need not be 'white', can serve as a tool to shift the system between various equivalence classes, i.e. as a kind of crosstalk and the source of a generalized stochastic resonance.

Deeply hidden in equation (24) is a very complicated pattern of equivalence class dynamics, since flows are defined on a manifold of languages having particular relations between H, S, and $\mathcal{M}$. Many possible information sources may, in fact, correspond to any particular 'point' on this manifold. Although we cannot pursue this in detail, as it involves subtle matters of 'topological groupoids' and the like, some implications are clear. In particular, setting equation (24) to zero and solving for 'stationary points' find a set of stable attractors, since the noise terms will perturb the structure from unstable equilibria. Second, what is converged to is not some 'stable state' in any sense, but rather is an equivalence class of highly dynamic information sources. We will have more to say on this below.

Convergence to more complicated structures, for example limit cycles or fractal 'strange attractors', is also possible in this model.

We have defined a particular set of equivalence classes of information sources dual to cognitive processes, ecosystems, and genetic heritage. That set parsimoniously characterizes the available dynamical manifolds, and, breaking of the associated groupoid symmetry creates more complex objects of considerable interest. This leads to the possibility, indeed, the necessity, of *Deus ex Machina* mechanisms to force transitions between the different possible modes within and across dynamic manifolds.

Equivalence classes of *states* gave dual information sources to cognitive systems. Equivalence classes of *information sources* give different characteristic system dynamics. Below we will examine equivalence classes of *paths*, which will produce different directed homotopy topologies characterizing those dynamical manifolds. This introduces the possibility of having different quasi-stable resilience modes *within* individual dynamic manifolds. Pink or white noise might provide a tunable means of creating crosstalk between different topological states within a dynamical manifold, or between different dynamical manifolds altogether.

Effectively, topological shifts between and within dynamic manifolds constitute a theory of phase transitions for information systems. Indeed, similar considerations have become central in the study of phase changes for physical systems. Franzosi and Pettini [37] and Pettini [68, 69] argue that the standard way of studying phase transition in physical systems is to consider how the values of thermodynamic observables, obtained in laboratory experiments, vary with temperature, volume, or an external field, and then to associate the experimentally observed discontinuities at a phase transition to the appearance of some kind of singularity entailing a loss of analyticity. However, they wonder whether this is the ultimate level of mathematical understanding of phase transition phenomena, or if some reduction to a more basic level is possible. Their theorem says that nonanalyticity is the 'shadow' of a more fundamental phenomenon occurring in configuration space: *a topology change*. Their theorem means that a topology change in a particular energy manifold is a *necessary* condition for a phase transition to take place. The topology changes implied here are those described within the framework of Morse theory through Morse-theoretic attachment handles. The converse of their theorem is not true. There is not a one-to-one correspondence between phase transitions and topology changes, and an open problem is that of *sufficiency* conditions, that is to determine which kinds of topology changes can entail the appearance of a phase transition.

The phenomenological Onsager treatment would also be enriched by adoption of a Morse theory perspective on topological transitions, following Michel and Mozrzymas [61].

The next section introduces a further topological complication.

7 Directed Homotopy

To reiterate, we can define equivalence classes of states according to whether they can be linked by grammatical/syntactical high probability 'meaningful' paths, generating 'languages'. Above we developed equivalence classes of languages constituting dynamic manifolds. Next we ask the precisely complementary question regarding paths on dynamical manifolds: For any two particular given states, is there some sense in which we can define equivalence classes across the set of meaningful paths linking them?

This is of particular interest to second order hierarchical models which, in effect, describe a universality class tuning of the renormalization parameters characterizing the dancing, flowing, tunably punctuated accession to high order cognitive function.

A closely similar question is central to recent algebraic geometry approaches to concurrent, i.e. highly parallel, computing (e.g., [40, 41, 72]), which we adapt.

For the moment we restrict the analysis to a system characterized by two Morse-theoretic parameters, say w_1 and w_2, and consider the set of meaningful paths connecting two particular points, say a and b, in the two dimensional w-space plane of figure 1. The arguments surrounding equations (17), (18) and (23) suggests that there may be regions of fatal attraction and strong repulsion, Black holes and White holes, which can either trap or deflect the path of institutional or multitasking machine cognition.

Figures 1 and 2 show two possible configurations for a Black and a White hole, diagonal and cross-diagonal. If one requires path monotonicity – always increasing or remaining the same – then, following, e.g., [40], figs. 6,7, there are, intuitively, two direct ways, without switchbacks, that one can get from a to b in the diagonal geometry of figure 1, without crossing a Black or White hole, but there are three in the cross-diagonal structure of figure 2.

Elements of each 'way' can be transformed into each other by continuous deformation without crossing either the Black or White hole. Figure 1 has two additional possible monotonic ways, involving over/under switchbacks, which are not drawn. Relaxing the monotonicity requirement generates a plethora of other possibilities, e.g., loopings and backwards switchbacks. It is not clear under what circumstances such complex paths can be meaningful, a matter for further study.

These ways are the equivalence classes defining the topological structure of the two different w-spaces, analogs to the fundamental homotopy groups in spaces which admit of loops (e.g., [53]). The closed loops needed for classical homotopy theory are impossible for this kind of system because of the 'flow of time'

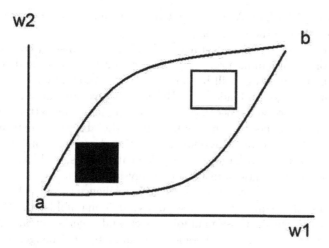

Fig. 1. Diagonal Black and White holes in the two dimensional w-plane. Only two direct paths can link points a and b which are continuously deformable into one another without crossing either hole. There are two additional monotonic switchback paths which are not drawn.

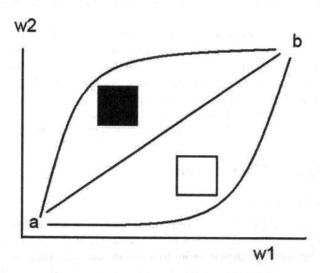

Fig. 2. Cross-diagonal Black and White holes. Three direct equivalence classes of continuously deformable paths can link a and b. Thus the two spaces are topologically distinct. Here monotonic switchbacks are not possible, although relaxation of that condition can lead to 'backwards' switchbacks and intermediate loopings.

defining the output of an information source – one goes from a to b, although, for nonmonotonic paths, intermediate looping would seem possible. The theory is thus one of directed homotopy, dihomotopy, and the central question revolves around the continuous deformation of paths in w-space into one another, without crossing Black or White holes. Goubault and Rausssen [41] provide another introduction to the formalism.

It seems likely that cultural heritage or developmental history can define quite different dihomotopies in natural ecosystems, cognitive process, and genetic heritage. That is, the topology will be developmentally modulated.

Such considerations, and indeed the Black Hole development of equation (23), suggest that a system which becomes trapped in a particular pattern of behavior cannot, in general, expect to emerge from it in the absence of external forcing mechanisms or the stochastic resonance/mutational action of 'noise'. Emerging from such a trap involves large-scale topological changes, and this is the functional equivalent of a phase transition in a physical system.

This sort of behavior is central to ecosystem resilience theory [44, 47, 48]. The essential idea is that equivalence classes of dynamic manifolds, and the directed homotopy classes within those manifolds, each and together create domains of quasi-stability requiring action of some external factor for change.

Apparently the set of dynamic manifolds, and its subsets of directed homotopy equivalence classes, formally classifies quasi-equilibrium states, and thus characterizes the different possible resilience modes.

Transitions between markedly different topological modes appear to be necessary effects of phase transitions, involving analogs to phase changes in physical systems.

Equivalence classes of quasi-languages generated dynamical manifolds, which [39, 84] use to construct a groupoid structure, and equivalence classes of paths on those manifolds constitute dihomotopy topological states. Shifts between dihomotopy modes represent transitions within manifolds, but larger scale shifts, between manifolds, are also possible, in this model.

Next we consider a particular canonical form of interaction between rapid, mesoscale, and slow information sources, which will produce the principal results.

8   Red Queen Coevolution

8.1   The Basic Idea

Natural systems subject to coevolutionary interaction may become enmeshed in the Red Queen dilemma of Alice in Wonderland, in that they must undergo constant evolutionary change in order to avoid extinction – they must constantly run just to stay in the same place. An example would be a competitive arms race between predator and prey: Each evolutionary advance in predation must be met with a coevolutionary adaptation which allows the prey to avoid the more efficient predator. Otherwise the system will become extinct, since a highly specialized predator can literally eat itself to extinction. Similarly, each prey defense must be matched by a predator adaptation for the system to persist.

Here we present a fairly elaborate model of coevolution, in terms of interacting information sources. Interaction events, we will argue, can be highly punctuated. These may be between Darwinian genetic, broadly cognitive, or embedding ecosystem structures.

We begin by examining ergodic information sources and their dynamics under the self-similarity of a renormalization transformation near a punctuated phase transition. We then study the linked interaction of two information sources in which the richness of the quasi-language of each affects the other, that is, when two information sources have become one another's primary environments. This leads directly and naturally to a coevolutionary Red Queen. We will generalize the development to a 'block diagonal' structure of several interacting sources, and extend the model to second order, producing a 'farming' metaphor.

The structures of interest to us here can be most weakly, and hence universally, described in terms of an adiabatically, piecewise stationary, ergodic information source involving a stochastic variate X which, in some general sense, sends symbols α in correlated sequences $\alpha_0, \alpha_1 ... \alpha_{n-1}$ of length n (which may vary), according to a joint probability distribution, and its associated conditional probability distribution,

$$P[X_0 = \alpha_0, X_1 = \alpha_1, ... X_{n-1} = \alpha_{n-1}],$$

$$P[X_{n-1} = \alpha_{n-1} | X_0 = \alpha_0, ... X_{n-2} = \alpha_{n-2}].$$

If the conditional probability distribution depends only on m previous values of X, then the information source is said to be of order m [5].

By 'ergodic' we mean that, in the long term, correlated sequences of symbols are generated at an average rate equal to their (joint) probabilities. 'Adiabatic' means that changes are slow enough to allow the necessary limit theorems to function, 'stationary' means that, between pieces, probabilities don't change (much), and 'piecewise' means that these properties hold between phase transitions, which are described using renormalization methods.

As the length of the (correlated) sequences increases without limit, the Shannon-McMillan Theorem permits division of all possible streams of symbols into two groups, a relatively small number characterized as meaningful, whose long-time behavior matches the underlying probability distribution, and an increasingly large set of gibberish with vanishingly small probability. Let $N(n)$ be the number of possible meaningful sequences of length n emitted by the source $\mathbf{X}$. Again, uncertainty of the source, $H[\mathbf{X}]$, can be defined by the subadditive relation

$$H[\mathbf{X}] = \lim_{n \to \infty} \frac{\log[N(n)]}{n}.$$

The Shannon-McMillan Theorem shows how to characterize $H[\mathbf{X}]$ directly in terms of the joint probability distribution of the source $\mathbf{X}$: $H[\mathbf{X}]$ is observable and can be calculated from the inferred pattern of joint probabilities.

Let $P[x_i | y_j]$ be the conditional probability that stochastic variate $X = x_i$ given that stochastic variate $Y = y_j$ and let $P[x_i, y_j]$ be the joint probability that $X = x_i$ and $Y = y_j$. Then the joint and conditional uncertainties of X and Y, $H(X, Y)$, and $H(X|Y)$ are given by expressions like those of equation (4).

And again, the Shannon-McMillan Theorem of states that the subadditive function for $H[\mathbf{X}]$ is given by the limits of equation (5).

Estimating the probabilities of the sequences $\alpha_0, ...\alpha_{n-1}$ from observation, the ergodic property allows us to use them to estimate the uncertainty of the source, i.e. of the behavioral language $\mathbf{X}$. That is, $H[\mathbf{X}]$ is directly measurable.

Some elementary consideration (e.g., [5, 21]) shows that source uncertainty has a least upper bound, a supremum, defined by the capacity of the channel along which information is transmitted. That is, there exists a number C defined by externalities such that $H[\mathbf{X}] \leq C$.

C is the maximum rate at which the external world can transmit information originating with the information source, or that internal workspaces can communicate. Much of the subsequent development could, in fact, be expressed using this relation.

Again recall the relation between the subadditive expression for source uncertainty and the free energy density of a physical system, as expressed by equation (12), which undergoes a phase transition depending on an inverse temperature parameter $K = 1/T$ at a critical temperature T_C.

Imposition of a renormalization symmetry on $F(K)$ in equation (12) describes, in the infinite volume limit, the behavior of the system at the phase transition in terms of scaling laws [95]. After some development, taking the limit $n \to \infty$ as an analog to the infinite volume limit of a physical system, we will apply this approach to a parametized source uncertainty. We will examine changes in structure as a fundamental 'inverse temperature' changes across the underlying system.

We use three parameters to describe the relations between an information source and its environment or between different interacting sources.

The first, $J \geq 0$, measures the degree to which acquired characteristics are transmitted. For systems without memory $J = 0$. $J \approx 0$ thus represents a high degree of genetic as opposed to cultural inheritance.

J will always remain distinguished, a kind of inherent direction or external field strength in the sense of [95].

The second parameter, $Q = 1/\mathcal{C} \geq 0$, represents the inverse availability of resources. $Q \approx 0$ thus represents a high ability to renew and maintain an particular enterprise, in a large sense.

The third parameter, $K = 1/T$, is an inverse index of a generalized temperature T, which we will more directly specify below in terms of the richness of interacting information sources.

We suppose further that the structure of interest is implicitly embedded in, and operates within the context of, a larger manifold stratified by metric distances.

Take these as multidimensional vector quantities $\mathbf{A}, \mathbf{B}, \mathbf{C}....$ $\mathbf{A}$ may represent location in space, time delay, or the like, and $\mathbf{B}$ may be determined through multivariate analysis of a spectrum of observed behavioral or other factors, in the largest sense, etc.

It may be possible to reduce the effects of these vectors to a function of their magnitudes $a = |\mathbf{A}|$, $b = |\mathbf{B}|$ and $c = |\mathbf{C}|$, etc. Define the *generalized distance* r as

$$r^2 = a^2 + b^2 + c^2 + ... \qquad (25)$$

To be more explicit, we assume an ergodic information source $\mathbf{X}$ is associated with the reproduction and/or persistence of a population, ecosystem, cognitive dual language or other structure. The source $\mathbf{X}$, its uncertainty $H[J, K, Q, \mathbf{X}]$ and its parameters J, K, Q all are assumed to depend *implicitly* on the embedding manifold, in particular on the metric r of equation (25).

A particularly elegant and natural formalism for generating such punctuation in our context involves application of Wilson's [95] program of renormalization symmetry – invariance under the renormalization transform – to source uncertainty defined on the r-manifold. The results predict that language in the most general sense, which includes the transfer of information within a a cognitive enterprise, or between an enterprise and an embedding context, will undergo sudden changes in structure analogous to phase transitions in physical systems. The view is complementary to recent analyses of sudden fragmentation in social networks, seen from the perspective of percolation theory.

We must, however, emphasize that this approach is argument by abduction from physical theory: Much current development surrounding self-organizing physical phenomena is based on the assumption that at phase transition a system looks the same under renormalization. That is, phase transition represents a stationary point for a renormalization transform in the sense that the transformed quantities are related by simple scaling laws to the original values.

Renormalization is a clustering semigroup transformation in which individual components of a system are combined according to a particular set of rules into a 'clumped' system whose behavior is a simplified average of those components. Since such clumping is a many-to-one condensation, there can be no unique inverse renormalization, and, as the Appendix shows, many possible forms of condensation.

Assume it possible to redefine characteristics of the information source $\mathbf{X}$ and J, K, Q as functions of averages across the manifold having metric r, which we write as R. That is, 'renormalize' by clustering the entire system in terms of blocks of different sized R.

Let $N(K, J, Q, n)$ be the number of high probability meaningful correlated sequences of length n *across the entire community* in the r-manifold of equation (25), given parameter values K, J, Q. We study changes in

$$H[K, J, Q, \mathbf{X}] \equiv \lim_{n \to \infty} \frac{\log[N(K, J, Q, n)]}{n}$$

as $K \to K_C$ and/or $Q \to Q_C$ for critical values K_C, Q_C at which the system begins to undergo a marked transformation from one kind of structure to another.

Given the metric of equation (25), a *correlation length*, $\chi(K, J, Q)$, can be defined as the average length in r-space over which structures involving a particular phase dominate.

Now clump the 'community' into blocks of average size R in the multivariate r-manifold, the 'space' in which the system of interest is implicitly embedded.

Following the classic argument of [95], reproduced and expanded in the Appendix, it is possible to impose renormalization symmetry on the source uncertainty on H and χ by assuming at transition the relations

$$H[K_R, J_R, Q_R, \mathbf{X}] = R^D H[K, J, Q, \mathbf{X}] \tag{26}$$

and

$$\chi(K_R, J_R, Q_R) = \frac{\chi(K, J, Q)}{R} \tag{27}$$

hold, where K_R, J_R and Q_R are the transformed values of K, J and Q after the clumping of renormalization. We take $K_1, J_1, Q_1 \equiv K, J, Q$ and permit the characteristic exponent D to be nonintegral. The Mathematical Appendix provides examples of other possible relations.

Equations (26) and (27) are assumed to hold in a neighborhood of the transition values K_C and Q_C.

Differentiating these with respect to R gives complicated expressions for $dK_R/dR, dJ_R/dR$ and dQ_R/dR depending simply on R which we write as

$$dK_R/dR = \frac{u(K_R, J_R, Q_R)}{R} \tag{28}$$

$$dQ_R/dR = \frac{w(K_R, J_R, Q_R)}{R}$$

$$dJ_R/dR = \frac{v(K_R, J_R, Q_R)}{R} J_R.$$

Solving these differential equations gives K_R, J_R and Q_R as functions of J, K, Q and R.

Substituting back into equations (26) and (27) and expanding in a first order Taylor series near the critical values K_C and Q_C gives power laws much like the Widom-Kadanoff relations for physical systems [95]. For example, letting $J = Q = 0$ and taking $\kappa \equiv (K_C - K)/K_C$ gives, in first order near K_C,

$$H = \kappa^{D/y} H_0 \tag{29}$$

$$\chi = \kappa^{-1/y} \chi_0$$

where y is a constant arising from the series expansion.

Note that there are only two fundamental equations – (26) and (27) – in $n > 2$ unknowns: The critical 'point' is, in this formulation, most likely to be a complicated implicitly defined critical surface in $J, K, Q, ...$-space. The 'external field strength' J remains distinguished in this treatment, i.e. the inverse of the degree to which acquired characteristics are inherited, but *neither K, Q nor other parameters are, by themselves, fundamental*, rather their joint interaction defines critical behavior along this surface.

That surface is a fundamental object, not the particular set of parameters (except for J) used to define it, which may be subject to any set of transformations which leave the surface invariant. Thus inverse generalized temperature resource availability or whatever other parameters may be identified as affecting the richness of cognition, are inextricably intertwined and mutually interacting, according to the form of this critical evolutionary transition surface. That surface, in turn, is unlikely to remain fixed, and should vary with time or other extrinsic parameters, including, but not likely limited to, J.

At the critical surface a Taylor expansion of the renormalization equations (26) and (27) gives a first order matrix of derivatives whose eigenstructure defines fundamental system behavior. For physical systems the surface is a saddle point [95], but more complicated behavior seems likely in what we study. See Binney et al., [12] for some details of this differential geometry.

Taking, for the moment, the simplest formulation, $(J = Q = 0)$, that is, a well-capitalized structure with memory, as K increases toward a threshold value K_C, the source uncertainty of the reproductive, behavioral or other language common across the community declines and, at K_C, the average regime dominated by the 'other phase' grows. That is, the system begins to freeze into one having a large correlation length for the second phase. The two phenomena are linked at criticality in physical systems by the scaling exponent y.

Assume the rate of change of $\kappa = (K_C - K)/K_C$ remains constant, $|d\kappa/dt| = 1/\tau_K$. Analogs with physical theory suggest there is a characteristic time constant for the phase transition, $\tau \equiv \tau_0/\kappa$, such that if changes in κ take place on a timescale longer than τ for any given κ, we may expect the correlation length $\chi = \chi_0 \kappa^{-s}$, $s = 1/y$, will be in equilibrium with internal changes and result in a very large fragment in r-space. Following Zurek [98, 99], the 'critical' freezout time, $\hat{t}$, will occur at a 'system time' $\hat{t} = \chi/|d\chi/dt|$ such that $\hat{t} = \tau$. Taking the derivative $d\chi/dt$, remembering that by definition $d\kappa/dt = 1/\tau_K$, gives

$$\frac{\chi}{|d\chi/dt|} = \frac{\kappa \tau_K}{s} = \frac{\tau_0}{\kappa},$$

so that

$$\kappa = \sqrt{s\tau_0/\tau_K}.$$

Substituting this value of κ into the equation for correlation length, the expected size of fragments in r-space, $d(\hat{t})$, becomes

$$d \approx \chi_0 \left(\frac{\tau_K}{s\tau_0}\right)^{s/2}$$

with $s = 1/y > 0$. The more rapidly K approaches K_C the smaller is τ_K and the smaller and more numerous are the resulting r-space fragments. Thus rapid change produces small fragments more likely to risk extinction in a system dominated by economies of scale.

8.2   Recursive Interaction

Extending the theory above involves envisioning reciprocally interacting genetic, cognitive or ecosystem information sources as subject to a coevolutionary Red

Queen by treating their respective source uncertainties as recursively parametized by each other. That is, *assume the information sources are each other's primary environments*. These are, respectively, characterized by information sources $\mathbf{X}$ and $\mathbf{Y}$, whose uncertainties are parametized

[1] by measures of both inheritance and inverse resources – $J\,Q$ as above – and, most critically,

[2] by each others inverse uncertainties, $\mathcal{H}_X \equiv 1/H[\mathbf{X}]$ and $\mathcal{H}_Y \equiv 1/H[\mathbf{Y}]$, i.e.

$$H[\mathbf{X}] = H[Q, J, \mathcal{H}_Y, \mathbf{X}] \tag{30}$$
$$H[\mathbf{Y}] = H[Q, J, \mathcal{H}_X, \mathbf{Y}].$$

This is a recursive system having complex behaviors.

Assume a strongly heritable genetic system, i.e. $J = 0$, with fixed inverse resource base, Q, for which $H[\mathbf{X}]$ follows something like the lower graph in figure 3, a reverse S-shaped curve with $K \equiv \mathcal{H}_Y = 1/H[\mathbf{Y}]$, and similarly $H[\mathbf{Y}]$ depends on $\mathcal{H}_X$. That is, increase or decline in the source uncertainty of one system leads to increase or decline in the source uncertainty of the other, and vice versa. The richness of the two information sources is closely linked.

Start at the right of the lower graph for $H[\mathbf{X}]$ in figure 3, the source uncertainty of the first system, but to the left of the critical point K_C. Assume $H[\mathbf{Y}]$ increases so $\mathcal{H}_Y$ decreases, and thus $H[\mathbf{X}]$ increases, walking up the lower curve of figure 3 from the right: the richness of the first system's internal language increases – or the interaction between internal structures increases the richness of their dual cognitive information sources – they get smarter or faster or more poisonous, or their herd behavior becomes more sophisticated in the presence of a predator.

The increase of $H[\mathbf{X}]$ leads, in turn, to a decline in $\mathcal{H}_X$ and triggers an increase of $H[\mathbf{Y}]$, whose increase leads to a further increase of $H[\mathbf{X}]$ and vice versa: The Red Queen, taking the system from the right of figure 3 to the left, up the lower curve as the two systems mutually interact.

Now enlarge the scale of the argument, and consider the possibility of other interactions.

The upper graph of figure 3 represents the disorder

$$S = H[K, \mathbf{X}] - K dH[K, \mathbf{X}]/dK, K \equiv 1/H[\mathbf{Y}].$$

According to the dynamical manifold analysis, the peak in S represents a repulsive barrier for transition between high and low values of $H[\mathbf{X}]$. This leads to the expectation of *hysteresis*. That is, the two realms, to the left and right of the peak in S for figure 3, thus represent quasi-stable resilience modes, in this model.

8.3   Extending the Model

The model directly generalizes to multiple interacting information sources.

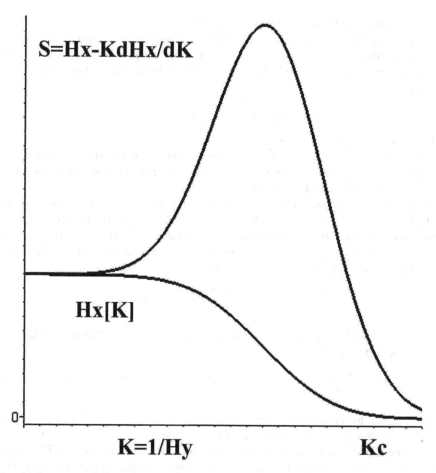

Fig. 3. A reverse-S-shaped curve for source uncertainty $H[\mathbf{X}]$ – measuring language richness – as a function of an inverse temperature parameter $K = 1/H[\mathbf{Y}]$. To the right of the critical point K_C the system breaks into fragments in r-space whose size is determined by the rate at which K approaches K_C. A collection of fragments already to the right of K_C, however, would be seen as condensing into a single unit as K declined below the critical point. If K is an inverse source uncertainty itself, i.e. $K = 1/H[\mathbf{Y}]$ for some information source $\mathbf{Y}$, then under such conditions a Red Queen dynamic can become enabled, driving the system strongly to the left. No intermediate points are asymptotically stable, given a genetic heritage in this development, although generalized Onsager/dynamical arguments suggest that the repulsive peak in $S = H - K/dH/dK$ can serve to create quasi-stable resilience realms. To the right of the critical point K_C the system is locked into disjoint fragments.

First consider a matrix of crosstalk measures between a set of information sources. Assume the matrix can be block diagonalized into two major components, characterized by network information source measures like equation (16),

$$I_m(X_1...X_i|Y_1...Y_j|Z_1...Z_k), m = 1, 2.$$

Then apply the two-component theory above.

Extending the development to multiple, recursively interacting information sources resulting from a more general block diagonalization seems direct. First use inverse measures $\mathcal{I}_j \equiv 1/I_j, j \neq m$ as parameters for each of the other blocks, writing

$$I_m = I_m(K_1...K_s,...\mathcal{I}_j...), j \neq m$$

where the K_s represent other relevant parameters.

Next segregate the $\mathcal{I}_j$ according to their relative rates of change, as in equation (16). Cognitive gene expression would be among the most rapid, followed by ecosystem dynamics and selection.

The dynamics of such a system, following the pattern of equations (18) and (24), becomes a recursive network of stochastic differential equations, similar to those used to study many other highly parallel dynamic structures (e.g., [96]).

Letting the K_j and $\mathcal{I}_m$ all be represented as parameters Q_j, (with the caveat that I_m not depend on $\mathcal{I}_m$), one can define

$$S_I^m \equiv I_m - \sum_i Q_i \partial I_m / \partial Q_i$$

to obtain a complicated recursive system of phenomenological 'Onsager relations' stochastic differential equations like (24),

$$dQ_t^j = \sum_i [L_{j,i}(t, ...\partial S_I^m/\partial Q^i...)dt + \sigma_{j,i}(t, ...\partial S_I^m/\partial Q^i...)dB_t^i], \qquad (31)$$

where, again, for notational simplicity only, we have expressed both the reciprocal $\mathcal{I}$'s and the external K's in terms of the same Q_j.

m ranges over the I_m and we could allow different kinds of 'noise' dB_t^i, having particular forms of quadratic variation which may, in fact, represent a projection of environmental factors under something like a rate distortion manifold [83].

Indeed, the I_m and/or the derived S^m might, in some circumstances, be combined into a Morse function, permitting application of Pettini's Topological Hypothesis.

The model rapidly becomes unwieldy, probably requiring some clever combinatorial or groupoid convolution algebra and related diagrams for concise expression, much as in the usual field theoretic perturbation expansions (Hopf algebras, for example). The virtual reaction method of [97] is another possible approach.

As in the simple model above, there will be, first, multiple quasi-stable points within a given system's I_m, representing a class of generalized resilience modes accessible via punctuation and enmeshing gene selection, gene expression, and ecological resilience – analogous to the simple model of figure 3.

Second, however, will be analogs to the fragmentation of figure 3 when the system exceeds the critical value K_c. That is, the K-parameter structure will represent full-scale fragmentation of the entire structure, and not just punctuation within it.

We thus infer two classes of punctuation possible for this kind of structure, both of which could entrain ecosystem resilience shifts, gene expression, and gene selection, although the latter kind would seem to be the far more dramatic.

There are other possible patterns: [1] Setting equation (31) equal to zero and solving for stationary points again gives attractor states since the noise terms preclude unstable equilibria. [2] Unlike equation (24), however, this system may converge to limit cycle or 'strange attractor' behaviors in which the system seems to chase its tail endlessly. [3] What is converged to in both cases is not a simple state or limit cycle of states. Rather it is an equivalence class, or set of them, of highly dynamic information sources coupled by mutual interaction through crosstalk. Thus 'stability' in this structure represents particular patterns of ongoing dynamics rather than some identifiable 'state'.

Here we are, at last and indeed, deeply enmeshed in a highly recursive phenomenological stochastic differential equations, but at a deeper level than Zhu et al. [97] envisioned for gene expression alone, and in a dynamic rather than static manner: the objects of this dynamical system are equivalence classes of information sources and their crosstalk, rather than simple 'states' of a dynamical or reactive chemical system.

Imposition of necessary conditions from the asymptotic limit theorems of communication theory has, at least in theory, beaten the thicket back one full layer.

Other formulations may well be possible, but our work here serves to illustrate the method.

It is, however, interesting to compare our results to those of Dieckmann and Law [24], who invoke evolutionary game dynamics to obtain a first order canonical equation having the form

$$ds_i/dt = K_i(s)\partial W_i(s_i', s)|_{s_i'=s_i}. \tag{32}$$

The s_i, with $i = 1, ..., N$ denote adaptive trait values in a community comprising N species. The $W_i(s_i', s)$ are measures of fitness of individuals with trait values s_i' in the environment determined by the resident trait values s, and the $K_i(s)$ are non-negative coefficients, possibly distinct for each species, that scale the rate of evolutionary change. Adaptive dynamics of this kind have frequently been postulated, based either on the notion of a hill-climbing process on an adaptive landscape or some other sort of plausibility argument.

When this equation is set equal to zero, so there is no time dependence, one obtains what are characterized as 'evolutionary singularities', i.e. stationary points.

Dieckmann and Law contend that their formal derivation of this equation satisfies four critical requirements: [1] The evolutionary process needs to be considered in a coevolutionary context. [2] A proper mathematical theory of evolution should be dynamical. [3] The coevolutionary dynamics ought to be underpinned by a microscopic theory. [4] The evolutionary process has important stochastic elements.

Our equation (31) seems clearly within this same ballpark, although we have taken a much different route, one which indeed produces elaborate patterns of phase transition punctuation in a highly natural manner. Champagnat et al. [16], in fact, derive a higher order canonical approximation extending equation (32) which is very much closer equation to (31), that is, a stochastic differential equation describing evolutionary dynamics. Champagnat et al. [16] go even further, using a large deviations argument to analyze dynamical coevolutionary paths, not merely evolutionary singularities. They contend that in general, the issue of evolutionary dynamics drifting away from trajectories predicted by the canonical equation can be investigated by considering the asymptotic of the probability of 'rare events' for the sample paths of the diffusion. By 'rare events' they mean diffusion paths drifting far away from the canonical equation. The probability of such rare events is governed by a large deviation principle: when a critical parameter (designated ϵ) goes to zero, the probability that the sample path of the diffusion is close to a given rare path ϕ decreases exponentially to 0 with rate $I(\phi)$, where the 'rate function' I can be expressed in terms of the parameters of the diffusion. This result, in their view, can be used to study long-time behavior of the diffusion process when there are multiple attractive evolutionary singularities. Under proper conditions the most likely path followed by the diffusion when exiting a basin of attraction is the one minimizing the rate function I over all the appropriate trajectories. The time needed to exit the basin is of the order $\exp(H/\epsilon)$ where H is a quasi-potential representing the minimum of the rate function I over all possible trajectories.

An essential fact of large deviations theory is that the rate function I which Champagnat et al. [16] invoke can almost always be expressed as a kind of entropy, that is, in the form $I = \sum_j P_j \log(P_j)$ for some probability distribution. This result goes under a number of names; Sanov's Theorem, Cramer's Theorem, the Gartner-Ellis Theorem, the Shannon-McMillan Theorem, and so forth (e.g., [22]). Here we will use it, in combination with the cognitive paradigm for gene expression, to suggest the possibility of second order effects in coevolutionary process. That is, gene expression, because of its underlying cognitive nature, may be an even more central aspect of coevolutionary process than is currently understood: The fluctuational paths defined by the system of equations in (31) may, under some conditions, become serially correlated outputs of an information source driven by cognitive gene expression. In particular, the coevolutionary pressures inherent to equation (31) may in fact strongly select for significant cognition in gene expression.

8.4   Second Order Theory: Farming a Coevolutionary System

We begin with a recapitulation of large deviations and fluctuation formalism.

Information source uncertainty, according to the Shannon-McMillan Theorem, serves as a splitting criterion between high and low probability sequences (or pairs of them) and displays the fundamental characteristic of a growing body of work in applied probability often termed the Large Deviations Program, (LDP) which seeks to unite information theory, statistical mechanics and the theory of fluctuations under a single umbrella.

Following [22], (p.2),

Let $X_1, X_2, ...X_n$ be a sequence of independent, standard Normal, real-valued random variables and let

$$S_n = \frac{1}{n} \sum_{j=1}^{n} X_j. \tag{33}$$

Since S_n is again a Normal random variable with zero mean and variance $1/n$, for all $\delta > 0$

$$\lim_{n \to \infty} P(|S_n| \geq \delta) = 0, \tag{34}$$

where P is the probability that the absolute value of S_n is greater or equal to δ. Some manipulation, however, gives

$$P(|S_n| \geq \delta) = 1 - \frac{1}{\sqrt{2\pi}} \int_{-\delta\sqrt{n}}^{\delta\sqrt{n}} \exp(-x^2/2)dx, \tag{35}$$

so that

$$\lim_{n \to \infty} \frac{\log P(|S_n| \geq \delta)}{n} = -\delta^2/2. \tag{36}$$

This can be rewritten for large n as

$$P(|S_n| \geq \delta) \approx \exp(-n\delta^2/2). \tag{37}$$

That is, for large n, the probability of a large deviation in S_n follows something much like the asymptotic equipartition relation of the Shannon-McMillan Theorem, i.e. that meaningful paths of length n all have approximately the same probability $P(n) \propto \exp(-nH[\mathbf{X}])$.

Questions about meaningful paths appear suddenly as formally isomorphic to the central argument of the LDP which encompasses statistical mechanics, fluctuation theory, and information theory into a single structure [22].

Perhaps the cardinal tenet of large deviation theory is that the rate function $-\delta^2/2$ can, under proper circumstances, be expressed as a mathematical entropy having the standard form

$$-\sum_{k} p_k \log p_k, \tag{38}$$

for some set of probabilities p_k. Again, this striking result goes under various names at various levels of approximation – Sanov's Theorem, Cramer's Theorem, the Gartner-Ellis Theorem, the Shannon-McMillan Theorem, and so on [22].

Next we briefly recapitulate part of the standard treatment of large fluctuations [38, 66].

The macroscopic behavior of a complicated physical system in time is assumed to be described by the phenomenological Onsager relations giving large-scale fluxes as

$$\sum_{i} R_{i,j} dK_j/dt = \partial S/\partial K_i, \tag{39}$$

where the $R_{i,j}$ are appropriate constants, S is the system entropy and the K_i are the generalized coordinates which parametize the system's free energy.

Entropy is defined from free energy F by a Legendre transform – more of which follows below:

$$S \equiv F - \sum_j K_j \partial F / \partial K_j,$$

where the K_j are appropriate system parameters.

Neglecting volume problems for the moment, free energy can be defined from the system's partition function Z as

$$F(K) = \log[Z(K)].$$

The partition function Z, in turn, is defined from the system Hamiltonian – defining the energy states – as

$$Z(K) = \sum_j \exp[-KE_j],$$

where K is an inverse temperature or other parameter and the E_j are the energy states.

Inverting the Onsager relations gives

$$dK_i/dt = \sum_j L_{i,j} \partial S / \partial K_j = L_i(K_1, ..., K_m, t) \equiv L_i(K, t). \tag{40}$$

The terms $\partial S / \partial K_i$ are macroscopic driving forces dependent on the entropy gradient.

Let a white Brownian noise $\epsilon(t)$ perturb the system, so that

$$dK_i/dt = \sum_j L_{i,j} \partial S / \partial K_j + \epsilon(t) \tag{41}$$

$$= L_i(K, t) + \epsilon(t),$$

where the time averages of ϵ are $< \epsilon(t) >= 0$ and $< \epsilon(t)\epsilon(0) >= D\delta(t)$. $\delta(t)$ is the Dirac delta function, and we take K as a vector in the K_i.

Following Luchinsky [59], if the probability that the system starts at some initial macroscopic parameter state K_0 at time $t = 0$ and gets to the state $K(t)$ at time t is $P(K, t)$, then a somewhat subtle development (e.g., [33]) gives the forward Fokker-Planck equation for P:

$$\partial P(K, t)/\partial t = -\nabla \cdot (L(K, t)P(K, t)) + (D/2)\nabla^2 P(K, t). \tag{42}$$

In the limit of weak noise intensity this can be solved using the WKB, i.e. the eikonal, approximation, as follows: take

$$P(K, t) = z(K, t) \exp(-s(K, t)/D). \tag{43}$$

$z(K, t)$ is a prefactor and $s(K, t)$ is a classical action satisfying the Hamilton-Jacobi equation, which can be solved by integrating the Hamiltonian equations of motion. The equation reexpresses $P(K,t)$ in the usual parametized negative exponential format.

Let $p \equiv \nabla s$. Substituting and collecting terms of similar order in D gives

$$dK/dt = p + L, dp/dt = -\partial L/\partial Kp \qquad (44)$$

and

$$-\partial s/\partial t \equiv h(K,p,t) = pL(K,t) + \frac{p^2}{2}, \qquad (45)$$

with $h(K,t)$ the Hamiltonian for appropriate boundary conditions.

Again following Luchinsky [59], these Hamiltonian equations have two different types of solution, depending on p. For $p = 0, dK/dt = L(K,t)$ which describes the system in the absence of noise. We expect that with finite noise intensity the system will give rise to a distribution about this deterministic path. Solutions for which $p \neq 0$ correspond to *optimal paths* along which the system will move with overwhelming probability.

These results can, however, again be directly derived as a special case of a Large Deviation Principle based on generalized entropies mathematically similar to Shannon's uncertainty from information theory, bypassing the Hamiltonian formulation entirely [22].

For a cognitive system characterized by a dual information source, of course, there is no Hamiltonian, but the generalized entropy or splitting criterion treatment still works. The trick is to do with information source uncertainty what is done here with a Hamiltonians.

Here we are concerned, not with a random Brownian distortion of simple physical systems, but, invoking cognitive gene expression, with a possibly complex behavioral structure, in the largest sense, composed of quasi-independent actors for which *meaningful/optimal paths have extremely structured serial correlation, amounting to a grammar and syntax, precisely the fact which allows definition of an information source* and enables the use of the very sparse equipartition of the Shannon-McMillan and Rate Distortion Theorems.

In sum, to again paraphrase [59], large fluctuations, although infrequent, are fundamental in a broad range of processes, and it was recognized by Onsager and Machlup [66] that insight into the problem could be gained from studying the distribution of fluctuational paths along which the system moves to a given state. This distribution is a fundamental characteristic of the fluctuational dynamics, and its understanding leads toward control of fluctuations. Fluctuational motion from the vicinity of a stable state may occur along different paths. For large fluctuations, the distribution of these paths peaks sharply along an optimal, most probable, path. In the theory of large fluctuations, the pattern of optimal paths plays a role similar to that of the phase portrait in nonlinear dynamics.

In this development meaningful paths driven by cognitive gene expression can play something of the role of optimal paths in the theory of large fluctuations which Champagnat et al. [16] have invoked, but without benefit of a Hamiltonian.

The spread of the possible spectrum of cognitive gene expression within a species, affecting the ability to adapt to changing ecological niches, then becomes central to the mitigation of selection pressures generated by coevolutionary dynamics: too limited a response repertoire will cause a species to become fully entrained into high probability dynamical fluctuational paths leading to punctuated extinction events. A broad spectrum allows a species to ride out much more of the coevolutionary selection pressure.

A sufficiently broad repertoire of cognitive gene expression responses leads, however, to the necessity of a second order coevolution model in which the high probability fluctuational paths defined by the system of equations (31) *are, in fact, themselves the output of some information source.* This is a model closely analogous to the second order cognitive structures needed to explain animal consciousness (e.g., [83, 93]). Intuitively, this transition to 'cognitive coevolution' would be particularly likely under the influence of a strong system of epigenetic inheritance, that is, an animal culture extending the niche spectrum offered by cognitive gene expression alone. Thus we could expand this development to one encompassing biocultural coevolution, in particular the development of agriculture, matters to be pursued in subsequent work.

9   Generalized Stochastic Resonance: Many Baldwin Effects

These arguments can be significantly extended using a tuning theorem variant of the Shannon Coding Theorem which leads to a generalized form of stochastic resonance applicable to a spectrum of phenomena similar to the Baldwin effect. Ancel [4] has presented what is perhaps the clearest mathematical model of the basic idea. She argues that organisms that make non-hereditary physical or behavioral modifications to survive environmental stresses will have better representation in future generations than less versatile organisms. Through natural selection then, the capacity for such adaptation along with the beneficial acquired traits will become universal. Calculation shows that the distribution of phenotypes in a population depends largely on the extent of environmental stochasticity. When the environment undergoes intermediate rates of fluctuation, the Simpson-Baldwin effect arises through the interaction of natural selection and mutation on norms of reaction. In a highly volatile environment by contrast, organisms benefit from plasticity, and consequently do not experience a Simpson-Baldwin channeling of phenotype possibility.

The essential point, from our perspective, is the importance of intermediate rates of environmental fluctuation, which mirrors the generalized stochastic resonance arguments we ultimately invoke.

In this section we reconsider such phenomena from a purely information theoretic perspective, recovering in the process something much like the 'no free lunch' theorem of computational optimization theory (e.g., English [31]).

Messages from an information source, seen as symbols x_j from some alphabet, each having probabilities P_j associated with a random variable X, are 'encoded' into the language of a 'transmission channel', a random variable Y with symbols y_k, having probabilities P_k, possibly with error. Someone receiving the symbol

y_k then retranslates it (without error) into some x_k, which may or may not be the same as the x_j that was sent.

More formally, the message sent along the channel is characterized by a random variable X having the distribution

$$P(X = x_j) = P_j, j = 1, ..., M.$$

The channel through which the message is sent is characterized by a second random variable Y having the distribution

$$P(Y = y_k) = P_k, k = 1, ..., L.$$

Let the joint probability distribution of X and Y be defined as

$$P(X = x_j, Y = y_k) = P(x_j, y_k) = P_{j,k}$$

and the conditional probability of Y given X as

$$P(Y = y_k | X = x_j) = P(y_k | x_j).$$

Again the Shannon uncertainty of X and Y independently and the joint uncertainty of X and Y together are defined respectively as

$$H(X) = - \sum_{j=1}^{M} P_j \log(P_j)$$

$$H(Y) = - \sum_{k=1}^{L} P_k \log(P_k)$$

$$H(X, Y) = - \sum_{j=1}^{M} \sum_{k=1}^{L} P_{j,k} \log(P_{j,k}).$$

The conditional uncertainty of Y given X is

$$H(Y|X) = - \sum_{j=1}^{M} \sum_{k=1}^{L} P_{j,k} \log[P(y_k | x_j)]$$

For any two stochastic variates X and Y, $H(Y) \geq H(Y|X)$, as knowledge of X generally gives some knowledge of Y. Equality occurs only in the case of stochastic independence.

Since $P(x_j, y_k) = P(x_j) P(y_k | x_j)$, we have $H(X|Y) = H(X, Y) - H(Y)$.

The information transmitted by translating the variable X into the channel transmission variable Y – possibly with error – and then retranslating without error the transmitted Y back into X is defined as $I(X|Y) \equiv H(X) - H(X|Y) = H(X) + H(Y) - H(X, Y)$.

See, for example, [5, 21, 51] for details. The essential point is that if there is no uncertainty in X given the channel Y, then there is no loss of information through transmission. In general this will not be true, and herein lies the essence of the theory.

Given a fixed vocabulary for the transmitted variable X, and a fixed vocabulary and probability distribution for the channel Y, we may vary the probability distribution of X in such a way as to maximize the information sent. The capacity of the channel is defined as

$$C \equiv \max_{P(X)} I(X|Y) \tag{46}$$

subject to the subsidiary condition that $\sum P(X) = 1$.

The critical trick of the Shannon Coding Theorem for sending a message with arbitrarily small error along the channel Y at any rate $R < C$ is to encode it in longer and longer 'typical' sequences of the variable X; that is, those sequences whose distribution of symbols approximates the probability distribution $P(X)$ above which maximizes C.

If $S(n)$ is the number of such 'typical' sequences of length n, then

$$\log[S(n)] \approx nH(X) \tag{47}$$

where $H(X)$ is the uncertainty of the stochastic variable defined above. Some consideration shows that $S(n)$ is much less than the total number of possible messages of length n. Thus, as $n \to \infty$, only a vanishingly small fraction of all possible messages is meaningful in this sense. This observation, after some considerable development, is what allows the Coding Theorem to work so well. In sum, the prescription is to encode messages in typical sequences, which are sent at very nearly the capacity of the channel. As the encoded messages become longer and longer, their maximum possible rate of transmission without error approaches channel capacity as a limit. Again, [5, 21, 51] provide details.

This approach can be, in a sense, inverted to give a tuning theorem which parsimoniously describes the essence of the Rate Distortion Manifold.

Telephone lines, optical wave guides and the tenuous plasma through which a planetary probe transmits data to earth may all be viewed in traditional information-theoretic terms as a *noisy channel* around which we must structure a message so as to attain an optimal error-free transmission rate.

Telephone lines, wave guides and interplanetary plasmas are, relatively speaking, fixed on the timescale of most messages, as are most sociogeographic networks. Indeed, the capacity of a channel, is defined by varying the probability distribution of the 'message' process X so as to maximize $I(X|Y)$.

Suppose there is some message X so critical that its probability distribution must remain fixed. The trick is to fix the distribution $P(x)$ but *modify the channel* – i.e. tune it – so as to maximize $I(X|Y)$. The *dual* channel capacity C^* can be defined as

$$C^* \equiv \max_{P(Y),P(Y|X)} I(X|Y) \tag{48}$$

But

$$C^* = \max_{P(Y),P(Y|X)} I(Y|X) \tag{49}$$

since
$$I(X|Y) = H(X) + H(Y) - H(X,Y) = I(Y|X).$$

Thus, in a purely formal mathematical sense, *the message transmits the channel*, and there will indeed be, according to the Coding Theorem, a channel distribution $P(Y)$ which maximizes C^*.

One may do better than this, however, by modifying the channel matrix $P(Y|X)$. Since

$$P(y_j) = \sum_{i=1}^{M} P(x_i)P(y_j|x_i),$$

$P(Y)$ is entirely defined by the channel matrix $P(Y|X)$ for fixed $P(X)$ and

$$C^* = \max_{P(Y),P(Y|X)} I(Y|X) = \max_{P(Y|X)} I(Y|X).$$

Calculating C^* requires maximizing the complicated expression

$$I(X|Y) = H(X) + H(Y) - H(X,Y)$$

which contains products of terms and their logs, subject to constraints that the sums of probabilities are 1 and each probability is itself between 0 and 1. Maximization is done by varying the channel matrix terms $P(y_j|x_i)$ within the constraints. This is a difficult problem in nonlinear optimization. However, for the special case $M = L$, C^* may be found by inspection:

If $M = L$, then choose

$$P(y_j|x_i) = \delta_{j,i} \tag{50}$$

where $\delta_{i,j}$ is 1 if $i = j$ and 0 otherwise. For this special case

$$C^* \equiv H(X)$$

with $P(y_k) = P(x_k)$ for all k. *Information is thus transmitted without error when the channel becomes 'typical' with respect to the fixed message distribution* $P(X)$.

If $M < L$ matters reduce to this case, but for $L < M$ information must be lost, leading to Rate Distortion limitations.

Thus modifying the channel may be a far more efficient means of ensuring transmission of an important message than encoding that message in a 'natural' language which maximizes the rate of transmission of information on a fixed channel.

We have examined the two limits in which either the distributions of $P(Y)$ or of $P(X)$ are kept fixed. The first provides the usual Shannon Coding Theorem, and the second a tuning theorem variant, i.e., a tunable, retina-like object we can call a Rate Distortion Manifold [39, 84]. It seems likely, however, than for many important systems $P(X)$ and $P(Y)$ will interpenetrate, to use Richard Lewontin's terminology. That is, $P(X)$ and $P(Y)$ will affect each other in characteristic ways, so that some form of mutual tuning may be the most effective strategy.

Classic stochastic resonance (SR) emerges from these arguments quite directly. The only coding possible under the conditions of SR is to add random noise to the amplitude of a structured signal which, by itself, is below threshold for triggering some powerful, highly nonlinear device. The only 'tuning' possible to random noise is to vary its amplitude. By the arguments above, there will be some optimum noise amplitude which will maximize the dual channel capacity, and hence the transmission rate of the signal via the powerful, threshold-driven, oscillator.

Similarly, in Ancel's model of the Simpson-Baldwin effect, the only 'tuning' possible the system, as she has presented it, is the extent of environmental stochasticity. By these arguments, then, there will be an 'optimal' level of stochasticity which drives the effect.

Clearly more complicated natural phenomena can be subject to analogous tuning, with particular sensitivity to 'intermediate' level processes. And thereupon hangs our tale.

10   Discussion and Conclusions

The basic point is the inevitability of punctuation in generalized coevolutionary interactions, representing fundamental structural changes in underlying manifolds, roughly analogous to the topological hypothesis of Pettini [68, 69]. Thus evolution, resilience, and cognitive phenomena, which can all be (at least crudely) represented by information sources, are inherently subject to punctuated equilibrium phenomena essentially similar to ecosystem resilience. This pattern will involve each individually, as well as their interactions, a consequence of the fundamental homology between information source uncertainty and free energy density.

Holling [48] finds, for ecosystems, an extended keystone hypothesis, that all ecosystems are controlled and organized by a small number of key plant, animal, and abiotic processes that structure the landscape at different scales. Similarly, he invokes an entrainment hypothesis, that within any one ecosystem, the periodicities and architectural attributes of the critical structuring processes will establish a nested set of periodicities and spatial features that become attractors for other variables. He argues that the degree to which small, fast events influence larger, slower ones is critically dependent upon mesoscale disturbance processes.

Our lowest common denominator information theoretic approach to coevolutionary interaction between genes, embedding ecosystem, and cognitive process identifies ecosystem phenomena as the driving mesoscale: cognitive phenomena are much faster, and (for large animals) genetic change much slower. The generalized stochastic resonance argument of the previous section provides a formal basis for these assertions.

That is, punctuated changes in ecosystem structure, the traditional purview of ecological resilience, appear able to entrain both Darwinian genetic and cognitive phenomena – including gene expression, triggering similarly punctuated outcomes, on top of the punctuation naturally inherent to these information systems.

Thus, while discontinuous phase transitions are 'natural' at all scales of biological information process, we argue here that punctuated changes in embedding ecosystem resilience regime will be particularly effective at entraining faster cognitive and slower Darwinian genetic structural phenomena. In particular, punctuated changes in ecosystem structure can write images of themselves onto genetic sequence structure in a punctuated manner, resulting in punctuated population extinction and/or speciation events on geologic timescales, and in sudden changes in gene expression and other cognitive phenomena on more rapid timescales.

This is not an entirely new idea. Laland et al. [52] have used a different methodology to reach similar conclusions. In their view there is increasing recognition that all organisms modify their environments through a process they characterize as 'niche construction'. Such modifications can have profound effects on the distribution and abundance of organisms, the influence of keystone species, the control of energy and material flows, residence and return times, ecosystem resilience, and specific trophic relationships. The consequences of environment modification by organisms, however, are not restricted to ecology, and organisms can affect both their own and each other's evolution by modifying sources of natural selection in their environments. They cite Lewontin's work, which points out that many of the activities of organisms, such as migration, hoarding of food resources, habitat selection, or thermoregulatory behavior, are adaptive precisely because they dampen statistical variation in the availability of environmental resources.

Laland et al. [52] argue that, hitherto, it has not been possible to apply evolutionary theory to ecosystems, because of the presence of nonevolving abiota in ecosystems. They suspect this obstacle has been largely responsible for preventing the full integration of ecosystem ecology with population-community ecology. However, in their view, adding the new process of niche construction to the established process of natural selection enables the incorporation of both abiotic environmental components and interactions among populations and abiota in ecosystems into evolutionary models an approach equally applicable to both population-community ecology and ecosystem-level ecology.

Somewhat earlier, Odling-Smee et al. [64] discussed these matters from the perspective of Lewontin, who has argued that the 'metaphor of adaptation' should be replaced by a 'metaphor of construction'. However, the acceptance of Lewontin's position, they state, demands more than just semantic adjustments to evolutionary theory. Niche construction changes the dynamic of the evolutionary process in fundamental ways because it precludes a description of evolutionary change relative only to autonomous environments. Instead, evolution now consists of endless cycles of natural selection and niche construction. Equally, it is no longer tenable from their perspective to assume that the only way organisms can contribute to evolutionary descent is by passing on fit or unfit genes to their descendants relative to their environments, because they can also pass on modifications in those environments that are better or worse suited to their genes. Adaptation becomes a two-way street in this theory.

More recently Dercole et al. [23] have addressed the problem using their version of equation (32) to produce very complex dynamical patterns, focusing on eco-evolutionary dynamics in communities contaiing 'slow' and 'fast' populations, which allows relaxation of the ecological equilibrium assumption.

Whitham [94], in parallel with our approach, takes a genetic framework associated with ecologically-dominant keystone species to examine what they call community and ecosystem phenotypes. They ask whether heritable traits in a single species can affect an entire ecosystem. Recent studies, they claim, show that such traits have predictable effects on community structure and ecosystem processes. Because these community and ecosystem phenotypes have a genetic basis and are heritable, they claim it is possible to apply the principles of population and quantitative genetics to place the study of complex communities and ecosystems within an evolutionary framework. This could, they assert, allow us to understand, for the first time, the genetic basis of ecosystem processes, and the effect of such phenomena as climate change and introduced transgenetic organisms on entire communities.

Whitham [94] goes on to define community evolution as a genetically based change in the ecological interactions that occur between species over time.

Here, by contrast, although we too focus on keystone scales, our particular innovation has been to reduce the dynamics of genetic inheritance, ecosystem persistence, and gene expression to a least common denominator as information sources operating at markedly different rates, but coupled by crosstalk into a broadly coevolutionary phenomenon marked at all scales by emergent 'phase transition' phenomena generating patterns of punctuated equilibrium.

We have, at times, grossly simplified the mathematical analysis. Invocation of equivalence class arguments leads naturally into deep groupoid structures and related topological generalizations, including Morse theory [84]. Taking a 'mean number' rather than the mean field approach of the Mathematical Appendix generates a qualitatively different class of exactly solvable models, based on giant component phase transitions in networks. Hybrids of the two are possible, and evolutionary process is unlikely to be at all constrained by formal mathematical tractability. In addition higher cognitive phenomena like individual or group consciousness require second order models analogous to hierarchical regression. Much of this is described in [84].

We conclude with E.C. Pielou's [70] important warning regarding ecological modeling:

> "...[Mathematical models] are easy to devise; even though the assumptions of which they are constructed may be hard to justify, the magic phrase 'let us assume that...' overrides objections temporarily. One is then confronted with a much harder task: How is such a model to be tested? The correspondence between a model's predictions and observed events is sometimes gratifyingly close but this cannot be taken to

imply the model's simplifying assumptions are reasonable in the sense that neglected complications are indeed negligible in their effects...

In my opinion the usefulness of models is great... [however] it consists *not in answer questions but in raising them.* Models can be used to inspire new field investigations and these are the only source of new knowledge as opposed to new speculation."

The principal model-based speculation of this work is that, via the mechanisms of Section 9, mesoscale ecosystem resilience shifts can entrain punctuated events of gene expression and other cognitive phenomena on more rapid time scales, and, in large part through such mechanisms of phenotype expression, slower genetic selection-induced changes, triggering punctuated equilibrium Darwinian evolutionary transitions on geologic time scales. The model we have invoked, unlike most related work, is a statistical one in which the asymptotic limit theorems of information theory impose necessary conditions on the behavior of ecosystems, Darwinian genetic selection, and gene expression. These necessary conditions, as the Central Limit Theorem does for regression theory, permit the construction of empirical models which can be fitted to data. Scientific inference is not in the model fitting itself, but rather, in the comparison of similar systems under different conditions, and the comparison of different systems under similar conditions. This semi-empirical approach is, perhaps, what most differentiates our developments from other attempts to model biological processes. We can, at best, impose necessary conditions through our formal development. The real science must then be done by real experiment.

It is worth noting that, for human populations in particular, several other layers of information sources, those of (Lamarckian) culture, and of individual and group consciousness and learning, become manifest, producing a rich stew of complicated and interesting phenomena [84, 91, 92].

11   Mathematical Appendix

11.1   The Shannon-McMillan Theorem

According to the structure of the underlying language of which a message is a particular expression, some messages are more 'meaningful' than others, that is, are in accord with the grammar and syntax of the language. The Shannon-McMillan or Asymptotic Equipartition Theorem, describes how messages themselves are to be classified.

Suppose a long sequence of symbols is chosen, using the output of the random variable X above, so that an output sequence of length n, with the form

$$x_n = (\alpha_0, \alpha_1, ..., \alpha_{n-1})$$

has joint and conditional probabilities

$$P(X_0 = \alpha_0, X_1 = \alpha_1, ..., X_{n-1} = \alpha_{n-1})$$

$$P(X_n = \alpha_n | X_0 = \alpha_0, ..., X_{n-1} = \alpha_{n-1}).$$

Using these probabilities we may calculate the conditional uncertainty

$$H(X_n | X_0, X_1, ..., X_{n-1}).$$

The uncertainty of the *information source*, $H[\mathbf{X}]$, is defined as

$$H[\mathbf{X}] \equiv \lim_{n \to \infty} H(X_n | X_0, X_1, ..., X_{n-1}). \tag{51}$$

In general

$$H(X_n | X_0, X_1, ..., X_{n-1}) \leq H(X_n).$$

Only if the random variables X_j are all stochastically independent does equality hold. If there is a maximum n such that, for all $m > 0$

$$H(X_{n+m} | X_0, ..., X_{n+m-1}) = H(X_n | X_0, ..., X_{n-1}),$$

then the source is said to be of *order* n. It is easy to show that

$$H[\mathbf{X}] = \lim_{n \to \infty} \frac{H(X_0, ... X_n)}{n+1}.$$

In general the outputs of the $X_j, j = 0, 1, ..., n$ are *dependent*. That is, the output of the communication process at step n depends on previous steps. Such serial correlation, in fact, is the very structure which enables most of what is done in this paper.

Here, however, the processes are all assumed stationary in time, that is, the serial correlations do not change in time, and the system is *stationary*.

A very broad class of such self-correlated, stationary, information sources, the so-called *ergodic* sources for which the long-run relative frequency of a sequence converges stochastically to the probability assigned to it, have a particularly interesting property:

It is possible, in the limit of large n, to divide all sequences of outputs of an ergodic information source into two distinct sets, S_1 and S_2, having, respectively, very high and very low probabilities of occurrence, with the source uncertainty providing the splitting criterion. In particular the Shannon-McMillan Theorem states that, for a (long) sequence having n (serially correlated) elements, the number of 'meaningful' sequences, $N(n)$ – those belonging to set S_1 – will satisfy the relation

$$\frac{\log[N(n)]}{n} \approx H[\mathbf{X}]. \tag{52}$$

More formally,

$$\lim_{n \to \infty} \frac{\log[N(n)]}{n} = H[\mathbf{X}] \tag{53}$$

$$= \lim_{n \to \infty} H(X_n | X_0, ..., X_{n-1})$$

$$= \lim_{n \to \infty} \frac{H(X_0, ..., X_n)}{n+1}.$$

Using the internal structures of the information source permits *limiting attention only to high probability 'meaningful' sequences of symbols.*

11.2   The Rate Distortion Theorem

The Shannon-McMillan Theorem can be expressed as the 'zero error limit' of the Rate Distortion Theorem [21, 22] which defines a splitting criterion that identifies high probability pairs of sequences. We follow closely the treatment of [21].

The origin of the problem is the question of representing one information source by a simpler one in such a way that the least information is lost. For example we might have a continuous variate between 0 and 100, and wish to represent it in terms of a small set of integers in a way that minimizes the inevitable distortion that process creates. Typically, for example, an analog audio signal will be replaced by a 'digital' one. The problem is to do this in a way which least distorts the *reconstructed* audio waveform.

Suppose the original stationary, ergodic information source Y with output from a particular alphabet generates sequences of the form

$$y^n = y_1, ..., y_n.$$

These are 'digitized,' in some sense, producing a chain of 'digitized values'

$$b^n = b_1, ..., b_n,$$

where the b-alphabet is much more restricted than the y-alphabet.

b^n is, in turn, *deterministically retranslated* into a reproduction of the original signal y^n. That is, each b^m is mapped on to a unique n-length y-sequence in the alphabet of the information source Y:

$$b^m \rightarrow \hat{y}^n = \hat{y}_1, ..., \hat{y}_n.$$

Note, however, that many y^n sequences may be mapped onto the *same* retranslation sequence $\hat{y}^n$, so that information will, in general, be lost.

The central problem is to explicitly minimize that loss.

The retranslation process defines a new stationary, ergodic information source, $\hat{Y}$.

The next step is to define a *distortion measure*, $d(y, \hat{y})$, which compares the original to the retranslated path. For example the *Hamming distortion* is

$$d(y, \hat{y}) = 1, y \neq \hat{y} \tag{54}$$
$$d(y, \hat{y}) = 0, y = \hat{y}.$$

For continuous variates the *Squared error distortion* is

$$d(y, \hat{y}) = (y - \hat{y})^2. \tag{55}$$

There are many possibilities.

The distortion between paths y^n and $\hat{y}^n$ is defined as

$$d(y^n, \hat{y}^n) = \frac{1}{n} \sum_{j=1}^{n} d(y_j, \hat{y}_j). \tag{56}$$

Suppose that with each path y^n and b^n-path retranslation into the y-language and denoted y^n, there are associated individual, joint, and conditional probability distributions

$$p(y^n), p(\hat{y}^n), p(y^n|\hat{y}^n).$$

The *average distortion* is defined as

$$D = \sum_{y^n} p(y^n) d(y^n, \hat{y}^n). \tag{57}$$

It is possible, using the distributions given above, to define the information transmitted from the incoming Y to the outgoing $\hat{Y}$ process in the usual manner, using the Shannon source uncertainty of the strings:

$$I(Y, \hat{Y}) \equiv H(Y) - H(Y|\hat{Y}) = H(Y) + H(\hat{Y}) - H(Y, \hat{Y}).$$

If there is no uncertainty in Y given the retranslation $\hat{Y}$, then no information is lost.

In general, this will not be true.

The *information rate distortion function* $R(D)$ for a source Y with a distortion measure $d(y, \hat{y})$ is defined as

$$R(D) = \min_{p(y,\hat{y}); \sum_{(y,\hat{y})} p(y)p(y|\hat{y})d(y,\hat{y}) \leq D} I(Y, \hat{Y}). \tag{58}$$

The minimization is over all conditional distributions $p(y|\hat{y})$ for which the joint distribution $p(y, \hat{y}) = p(y)p(y|\hat{y})$ satisfies the average distortion constraint (i.e., average distortion $\leq D$).

The *Rate Distortion Theorem* states that $R(D)$ *is the maximum achievable rate of information transmission which does not exceed the distortion* D. Cover and Thomas [21] or Dembo and Zeitouni [22] provide details.

More to the point, however, is the following: Pairs of sequences $(y^n, \hat{y}^n)$ can be defined as *distortion typical*; that is, for a given average distortion D, defined in terms of a particular measure, pairs of sequences can be divided into two sets, a high probability one containing a relatively small number of (matched) pairs with $d(y^n, \hat{y}^n) \leq D$, and a low probability one containing most pairs. As $n \to \infty$, the smaller set approaches unit probability, and, for those pairs,

$$p(y^n) \geq p(\hat{y}^n|y^n) \exp[-nI(Y, \hat{Y})]. \tag{59}$$

Thus, roughly speaking, $I(Y, \hat{Y})$ embodies the splitting criterion between high and low probability pairs of paths.

For the theory of interacting information sources, then, $I(Y, \hat{Y})$ can play the role of H in the dynamic treatment above.

The rate distortion function can actually be calculated in many cases by using a Lagrange multiplier method – see Section 13.7 of [21].

11.3   Morse Theory

Morse theory examines relations between analytic behavior of a function – the location and character of its critical points – and the underlying topology of the manifold on which the function is defined. We are interested in a number of such functions, for example information source uncertainty on a parameter space and 'second order' iterations involving parameter manifolds determining critical behavior, for example sudden onset of a giant component in the mean number model, and universality class tuning in the mean field model. These can be reformulated from a Morse theory perspective. Here we follow closely the elegant treatments of [50, 68, 69].

The essential idea of Morse theory is to examine an n-dimensional manifold M as decomposed into level sets of some function $f : M \to \mathbf{R}$ where $\mathbf{R}$ is the set of real numbers. The a-level set of f is defined as

$$f^{-1}(a) = \{x \in M : f(x) = a\},$$

the set of all points in M with $f(x) = a$. If M is compact, then the whole manifold can be decomposed into such slices in a canonical fashion between two limits, defined by the minimum and maximum of f on M. Let the part of M below a be defined as

$$M_a = f^{-1}(-\infty, a] = \{x \in M : f(x) \le a\}.$$

These sets describe the whole manifold as a varies between the minimum and maximum of f.

Morse functions are defined as a particular set of smooth functions $f : M \to \mathbf{R}$ as follows. Suppose a function f has a critical point x_c, so that the derivative $df(x_c) = 0$, with critical value $f(x_c)$. Then f is a Morse function if its critical points are nondegenerate in the sense that the Hessian matrix of second derivatives at x_c, whose elements, in terms of local coordinates are

$$H_{i,j} = \partial^2 f / \partial x^i \partial x^j,$$

has rank n, which means that it has only nonzero eigenvalues, so that there are no lines or surfaces of critical points and, ultimately, critical points are isolated.

The index of the critical point is the number of negative eigenvalues of H at x_c.

A level set $f^{-1}(a)$ of f is called a critical level if a is a critical value of f, that is, if there is at least one critical point $x_c \in f^{-1}(a)$.

Again following [68], the essential results of Morse theory are:

[1] If an interval $[a, b]$ contains no critical values of f, then the topology of $f^{-1}[a, v]$ does not change for any $v \in (a, b)$. Importantly, the result is valid even if f is not a Morse function, but only a smooth function.

[2] If the interval $[a, b]$ contains critical values, the topology of $f^{-1}[a, v]$ changes in a manner determined by the properties of the matrix H at the critical points.

[3] If $f : M \to \mathbf{R}$ is a Morse function, the set of all the critical points of f is a discrete subset of M, i.e. critical points are isolated. This is Sard's Theorem.

[4] If $f : M \to \mathbf{R}$ is a Morse function, with M compact, then on a finite interval $[a, b] \subset \mathbf{R}$, there is only a finite number of critical points p of f such that $f(p) \in [a, b]$. The set of critical values of f is a discrete set of $\mathbf{R}$.

[5] For any differentiable manifold M, the set of Morse functions on M is an open dense set in the set of real functions of M of differentiability class r for $0 \le r \le \infty$.

[6] Some topological invariants of M, that is, quantities that are the same for all the manifolds that have the same topology as M, can be estimated and sometimes computed exactly once all the critical points of f are known: Let the Morse numbers $\mu_i (i = 1, ..., m)$ of a function f on M be the number of critical points of f of index i, (the number of negative eigenvalues of H). The Euler characteristic of the complicated manifold M can be expressed as the alternating sum of the Morse numbers of any Morse function on M,

$$\chi = \sum_{i=0}^{m} (-1)^i \mu_i.$$

The Euler characteristic reduces, in the case of a simple polyhedron, to

$$\chi = V - E + F$$

where V, E, and F are the numbers of vertices, edges, and faces in the polyhedron.

[7] Another important theorem states that, if the interval $[a, b]$ contains a critical value of f with a single critical point x_c, then the topology of the set M_b defined above differs from that of M_a in a way which is determined by the index, i, of the critical point. Then M_b is homeomorphic to the manifold obtained from attaching to M_a an i-handle, i.e. the direct product of an i-disk and an $(m - i)$-disk.

Again, Pettini [68] contains both mathematical details and further references. See, for example, Matusmoto [60] or the classic by Milnor [62].

11.4   The Mean Field Model

Wallace and Wallace [85, 86] have addressed how a language, in a large sense, 'spoken' on a network structure, responds as properties of the network change. The language might be speech, pattern recognition, or cognition. The network might be social, chemical, or neural. The properties of interest were the magnitude of 'strong' or 'weak' ties which, respectively, either disjointly partitioned the network or linked it across such partitioning. These would be analogous to local and mean-field couplings in physical systems.

Fix the magnitude of strong ties – again, those which disjointly partition the underlying network into cognitive or other submodules – but vary the index of nondisjunctive weak ties, P, between components, taking $K = 1/P$.

Assume the piecewise, adiabatically stationary ergodic information source (or sources) dual to cognitive process depends on three parameters, two explicit and one implicit. The explicit are K as above and, as a calculational device, an 'external field strength' analog J, which gives a 'direction' to the system. We will, in the limit, set $J = 0$. Note that many other approaches may well be possible, since renormalization techniques are more philosophy than prescription.

The implicit parameter, r, is an inherent generalized 'length' characteristic of the phenomenon, on which J and K are defined. That is, J and K are written as functions of averages of the parameter r, which may be quite complex, having nothing at all to do with conventional ideas of space. For example r may be defined by the degree of niche partitioning in ecosystems or separation in social structures.

For a given generalized language of interest having a well defined (adiabatically, piecewise stationary) ergodic source uncertainty, $H = H[K, J, \mathbf{X}]$.

To summarize a long train of standard argument [12, 95], imposition of invariance of H under a renormalization transform in the implicit parameter r leads to expectation of both a critical point in K, written K_C, reflecting a phase transition to or from collective behavior across the entire array, and of power laws for system behavior near K_C. Addition of other parameters to the system results in a 'critical line' or surface.

Let $\kappa \equiv (K_C - K)/K_C$ and take χ as the 'correlation length' defining the average domain in r-space for which the information source is primarily dominated by 'strong' ties. The first step is to average across r-space in terms of 'clumps' of length $R = <r>$. Then $H[J, K, \mathbf{X}] \rightarrow H[J_R, K_R, \mathbf{X}]$.

Taking Wilson's [95] analysis as a starting point – not the only way to proceed – the 'renormalization relations' used here are:

$$H[K_R, J_R, \mathbf{X}] = f(R)H[K, J, \mathbf{X}] \tag{60}$$

$$\chi(K_R, J_R) = \frac{\chi(K, J)}{R},$$

with $f(1) = 1$ and $J_1 = J, K_1 = K$. The first equation significantly extends Wilson's treatment. It states that 'processing capacity,' as indexed by the source uncertainty of the system, representing the 'richness' of the generalized language, grows monotonically as $f(R)$, which must itself be a dimensionless function in R, since both $H[K_R, J_R]$ and $H[K, J]$ are themselves dimensionless. Most simply, this requires replacing R by R/R_0, where R_0 is the 'characteristic length' for the system over which renormalization procedures are reasonable, then setting $R_0 \equiv 1$, hence measuring length in units of R_0.

Wilson's original analysis focused on free energy density. Under 'clumping,' densities must remain the same, so that if $F[K_R, J_R]$ is the free energy of the clumped system, and $F[K, J]$ is the free energy density before clumping, then Wilson's equation (4) is $F[K, J] = R^{-3}F[K_R, J_R]$,

$$F[K_R, J_R] = R^3 F[K, J].$$

Remarkably, the renormalization equations are solvable for a broad class of functions $f(R)$, or more precisely, $f(R/R_0)$, $R_0 \equiv 1$.

The second equation just states that the correlation length simply scales as R.

Again, the central feature of renormalization in this context is the assumption that, at criticality, the system looks the same at all scales, that is, it is *invariant under renormalization* at the critical point. All else flows from this.

There is no unique renormalization procedure for information sources: other, very subtle, symmetry relations – not necessarily based on the elementary physical analog we use here – may well be possible. This is important, since biological or social systems may well alter their renormalization properties – equivalent to tuning their phase transition dynamics – in response to external signals. We will make much use of a simple version of this possibility, termed 'universality class tuning,' below.

To begin, following Wilson, take $f(R) = R^d$, d some real number $d > 0$, and restrict K to near the 'critical value' K_C. If $J \to 0$, a simple series expansion and some clever algebra [12, 95] gives

$$H = H_0 \kappa^\alpha \tag{61}$$
$$\chi = \frac{\chi_0}{\kappa^s},$$

where α, s are positive constants. More biologically relevant examples appear below.

Further from the critical point, matters are more complicated, appearing to involve Generalized Onsager Relations, 'dynamical groupoids', and a kind of thermodynamics associated with a Legendre transform of H: $S \equiv H - KdH/dK$. Although this extension is quite important to describing behaviors away from criticality, the mathematical detail is cumbersome. A more detailed discussion appears at the end of this appendix.

An essential insight is that *regardless of the particular renormalization properties, sudden critical point transition is possible in the opposite direction for this model.* That is, going from a number of independent, isolated and fragmented systems operating individually and more or less at random, into a single large, interlocked, coherent structure, once the parameter K, the inverse strength of weak ties, falls below threshold, or, conversely, once the strength of weak ties parameter $P = 1/K$ becomes large enough.

Thus, increasing nondisjunctive weak ties between them can bind several different cognitive 'language' functions into a single, embedding hierarchical metalanguage containing each as a linked subdialect, and do so in an inherently punctuated manner. This could be a dynamic process, creating a shifting, ever-changing pattern of linked cognitive submodules, according to the challenges or opportunities faced by the organism.

This heuristic insight can be made more exact using a rate distortion argument (or, more generally, using the Joint Asymptotic Equipartition Theorem) as follows:

Suppose that two ergodic information sources Y and B begin to interact, to 'talk' to each other, to influence each other in some way so that it is possible, for example, to look at the output of B – strings b – and infer something about the behavior of Y from it – strings y. We suppose it possible to define a retranslation from the B-language into the Y-language through a deterministic code book, and call $\hat{Y}$ the translated information source, as mirrored by B.

Define some distortion measure comparing paths y to paths $\hat{y}$, $d(y, \hat{y})$. Invoke the Rate Distortion Theorem's mutual information $I(Y, \hat{Y})$, which is the splitting criterion between high and low probability pairs of paths. Impose, now, a parametization by an inverse coupling strength K, and a renormalization representing the global structure of the system coupling. This may be much different from the renormalization behavior of the individual components. If $K < K_C$, where K_C is a critical point (or surface), the two information sources will be closely coupled enough to be characterized as condensed.

In the absence of a distortion measure, the Joint Asymptotic Equipartition Theorem gives a similar result.

Detailed coupling mechanisms will be sharply constrained through regularities of grammar and syntax imposed by limit theorems associated with phase transition.

Biological renormalization. Next the mathematical detail concealed by the invocation of the asymptotic limit theorems emerges with a vengeance. Equation (60) states that the information source and the correlation length, the degree of coherence on the underlying network, scale under renormalization clustering in chunks of size R as

$$H[K_R, J_R]/f(R) = H[J, K]$$

$$\chi[K_R, J_R]R = \chi(K, J),$$

with $f(1) = 1, K_1 = K, J_1 = J$, where we have slightly rearranged terms.

Differentiating these two equations with respect to R, so that the right hand sides are zero, and solving for dK_R/dR and dJ_R/dR gives, after some consolidation, expressions of the form

$$dK_R/dR = u_1 d\log(f)/dR + u_2/R \tag{62}$$

$$dJ_R/dR = v_1 J_R d\log(f)/dR + \frac{v_2}{R} J_R.$$

The $u_i, v_i, i = 1, 2$ are functions of K_R, J_R, but not explicitly of R itself.

We expand these equations about the critical value $K_R = K_C$ and about $J_R = 0$, obtaining

$$dK_R/dR = (K_R - K_C)yd\log(f)/dR + (K_R - K_C)z/R \qquad (63)$$
$$dJ_R/dR = wJ_Rd\log(f)/dR + xJ_R/R.$$

The terms $y = du_1/dK_R|_{K_R=K_C}, z = du_2/dK_R|_{K_R=K_C}, w = v_1(K_C,0), x = v_2(K_C,0)$ are constants.

Solving the first of these equations gives

$$K_R = K_C + (K - K_C)R^z f(R)^y, \qquad (64)$$

again remembering that $K_1 = K, J_1 = J, f(1) = 1$.

Wilson's essential trick is to iterate on this relation, which is supposed to converge rapidly near the critical point [12], assuming that for K_R near K_C, we have

$$K_C/2 \approx K_C + (K - K_C)R^z f(R)^y. \qquad (65)$$

We iterate in two steps, first solving this for $f(R)$ in terms of known values, and then solving for R, finding a value R_C that we then substitute into the first of equations (60) to obtain an expression for $H[K,0]$ in terms of known functions and parameter values.

The first step gives the general result

$$f(R_C) \approx \frac{[K_C/(K_C - K)]^{1/y}}{2^{1/y}R_C^{z/y}}. \qquad (66)$$

Solving this for R_C and substituting into the first expression of equation (55) gives, as a first iteration of a far more general procedure [78], the result

$$H[K,0] \approx \frac{H[K_C/2,0]}{f(R_C)} = \frac{H_0}{f(R_C)} \qquad (67)$$
$$\chi(K,0) \approx \chi(K_C/2,0)R_C = \chi_0 R_C,$$

which are the essential relationships.

Note that a power law of the form $f(R) = R^m, m = 3$, which is the direct physical analog, may not be biologically reasonable, since it says that 'language richness' can grow very rapidly as a function of increased network size. Such rapid growth is simply not observed.

Taking the biologically realistic example of non-integral 'fractal' exponential growth,

$$f(R) = R^\delta, \qquad (68)$$

where $\delta > 0$ is a real number which may be quite small, equation (66) can be solved for R_C, obtaining

$$R_C = \frac{[K_C/(K_C - K)]^{[1/(\delta y + z)]}}{2^{1/(\delta y + z)}} \qquad (69)$$

for K near K_C. Note that, for a given value of y, one might characterize the relation $\alpha \equiv \delta y + z = $ constant as a 'tunable universality class relation' in the sense of Albert and Barabasi [3].

Substituting this value for R_C back into equation (66) gives a complex expression for H, having three parameters: δ, y, z.

A more biologically interesting choice for $f(R)$ is a logarithmic curve that 'tops out', for example

$$f(R) = m\log(R) + 1. \tag{70}$$

Again $f(1) = 1$.

Using Mathematica 4.2 or above to solve equation (66) for R_C gives

$$R_C = [\frac{Q}{LambertW[Q\exp(z/my)]}]^{y/z}, \tag{71}$$

where

$$Q \equiv (z/my)2^{-1/y}[K_C/(K_C - K)]^{1/y}.$$

The transcendental function LambertW(x) is defined by the relation

$$LambertW(x)\exp(LambertW(x)) = x.$$

It arises in the theory of random networks and in renormalization strategies for quantum field theories.

An asymptotic relation for $f(R)$ would be of particular biological interest, implying that 'language richness' increases to a limiting value with population growth. Such a pattern is broadly consistent with calculations of the degree of allelic heterozygosity as a function of population size under a balance between genetic drift and neutral mutation [46, 75]. Taking

$$f(R) = \exp[m(R-1)/R] \tag{72}$$

gives a system which begins at 1 when $R = 1$, and approaches the asymptotic limit $\exp(m)$ as $R \to \infty$. Mathematica 4.2 finds

$$R_C = \frac{my/z}{LambertW[A]}, \tag{73}$$

where

$$A \equiv (my/z)\exp(my/z)[2^{1/y}[K_C/(K_C - K)]^{-1/y}]^{y/z}.$$

These developments indicate the possibility of taking the theory significantly beyond arguments by abduction from simple physical models, although the notorious difficulty of implementing information theory existence arguments will undoubtedly persist.

Universality class distribution. Physical systems undergoing phase transition usually have relatively pure renormalization properties, with quite different systems clumped into the same 'universality class,' having fixed exponents at transition [12]. Biological and social phenomena may be far more complicated:

If the system of interest is a mix of subgroups with different values of some significant renormalization parameter m in the expression for $f(R,m)$, according to a distribution $\rho(m)$, then the first expression in equation (60) should generalize, at least to first order, as

$$H[K_R, J_R] = < f(R, m) > H[K, J] \tag{74}$$

$$\equiv H[K, J] \int f(R, m)\rho(m)dm.$$

If $f(R) = 1 + m\log(R)$ then, given any distribution for m,

$$< f(R) >= 1+ < m > \log(R) \tag{75}$$

where $< m >$ is simply the mean of m over that distribution.

Other forms of $f(R)$ having more complicated dependencies on the distributed parameter or parameters, like the power law R^δ, do not produce such a simple result. Taking $\rho(\delta)$ as a normal distribution, for example, gives

$$< R^\delta >= R^{<\delta>} \exp[(1/2)(\log(R^\sigma))^2], \tag{76}$$

where σ^2 is the distribution variance. The renormalization properties of this function can be determined from equation (66), and the calculation is left to the reader as an exercise, best done in Mathematica 4.2 or above.

Thus the information dynamic phase transition properties of mixed systems will not in general be simply related to those of a single subcomponent, a matter of possible empirical importance: If sets of relevant parameters defining renormalization universality classes are indeed distributed, experiments observing pure phase changes may be very difficult. Tuning among different possible renormalization strategies in response to external signals would result in even greater ambiguity in recognizing and classifying information dynamic phase transitions.

Important aspects of mechanism may be reflected in the combination of renormalization properties and the details of their distribution across subsystems.

In sum, real biological, social, or interacting biopsychosocial systems are likely to have very rich patterns of phase transition which may not display the simplistic, indeed, literally elemental, purity familiar to physicists. Overall mechanisms will, however, still remain significantly constrained by the theory, in the general sense of probability limit theorems.

Punctuated universality class tuning. The next step is to iterate the general argument onto the process of phase transition itself, producing a model of consciousness as a tunable neural workspace subject to inherent punctuated detection of external events.

As described above, an essential character of physical systems subject to phase transition is that they belong to particular 'universality classes'. Again, this means that the exponents of power laws describing behavior at phase transition will be the same for large groups of markedly different systems, with 'natural' aggregations representing fundamental class properties [12].

It appears that biological or social systems undergoing phase transition analogs need not be constrained to such classes, and that 'universality class tuning', meaning the strategic alteration of parameters characterizing the renormalization properties of punctuation, might well be possible. Here we focus on the tuning of parameters within a single, given, renormalization relation. Clearly, however, wholesale shifts of renormalization properties must ultimately be considered as well, a matter for future work.

Universality class tuning has been observed in models of 'real world' networks. As Albert and Barabasi [3] put it,

> "The inseparability of the topology and dynamics of evolving networks is shown by the fact that [the exponents defining universality class] are related by [a] scaling relation..., underlying the fact that a network's assembly uniquely determines its topology. However, in no case are these exponents unique. They can be tuned continuously..."

Suppose that a structured external environment, itself an appropriately regular information source $\mathbf{Y}$, 'engages' a modifiable cognitive system. The environment begins to write an image of itself on the cognitive system in a distorted manner permitting definition of a mutual information $I[K]$ splitting criterion according to the Rate Distortion or Joint Asymptotic Equipartition Theorems. K is an inverse coupling parameter between system and environment. At punctuation – near some critical point K_C – the systems begin to interact very strongly indeed, and, near K_C, using the simple physical model of equation (61),

$$I[K] \approx I_0 [\frac{K_C - K}{K_C}]^\alpha.$$

For a physical system α is fixed, determined by the underlying 'universality class.' Here we will allow α to vary, and, in the section below, to itself respond explicitly to signals.

Normalizing K_C and I_0 to 1,

$$I[K] \approx (1 - K)^\alpha. \tag{77}$$

The horizontal line $I[K] = 1$ corresponds to $\alpha = 0$, while $\alpha = 1$ gives a declining straight line with unit slope which passes through 0 at $K = 1$. Consideration shows there are progressively sharper transitions between the necessary zero value at $K = 1$ and the values defined by this relation for $0 < K, \alpha < 1$. The rapidly rising slope of transition with declining α is of considerable significance:

The instability associated with the splitting criterion $I[K]$ is defined by

$$Q[K] \equiv -KdI[K]/dK = \alpha K(1 - K)^{\alpha - 1}, \tag{78}$$

and is singular at $K = K_C = 1$ for $0 < \alpha < 1$. We interpret this to mean that values of $0 < \alpha \ll 1$ are highly unlikely for real systems, since $Q[K]$, in this model, represents a kind of barrier for 'social' information systems, in particular interacting neural network modules, a matter explored further below.

On the other hand, smaller values of α mean that the system is far more efficient at responding to the adaptive demands imposed by the embedding structured environment, since the mutual information which tracks the matching of internal response to external demands, $I[K]$, rises more and more quickly toward the maximum for smaller and smaller α as the inverse coupling parameter K declines below $K_C = 1$. That is, systems able to attain smaller α are more responsive to external signals than those characterized by larger values, in this model, but smaller values will be harder to reach, probably only at some considerable physiological or opportunity cost. Focused conscious action takes resources, of one form or another.

Wallace [92] makes these considerations explicit, modeling the role of contextual and energy constraints on the relations between Q, I, and other system properties.

The more biologically realistic renormalization strategies given above produce sets of several parameters defining the universality class, whose tuning gives behavior much like that of α in this simple example.

Formal iteration of the phase transition argument on this calculation gives tunable consciousness, focusing on paths of universality class parameters.

Suppose the renormalization properties of a language-on-a network system at some 'time' k are characterized by a set of parameters $A_k \equiv \alpha_1^k, ..., \alpha_m^k$. Fixed parameter values define a particular universality class for the renormalization. We suppose that, over a sequence of 'times,' the universality class properties can be characterized by a path $x_n = A_0, A_1, ..., A_{n-1}$ having significant serial correlations which, in fact, permit definition of an adiabatically piecewise stationary ergodic information source associated with the paths x_n. We call that source $\mathbf{X}$.

Suppose also, in the now-usual manner, that the set of external (or internal, systemic) signals impinging on consciousness is also highly structured and forms another information source $\mathbf{Y}$ which interacts not only with the system of interest globally, but specifically with its universality class properties as characterized by $\mathbf{X}$. $\mathbf{Y}$ is necessarily associated with a set of paths y_n.

Pair the two sets of paths into a joint path, $z_n \equiv (x_n, y_y)$ and invoke an inverse coupling parameter, K, between the information sources and their paths. This leads, by the arguments above, to phase transition punctuation of $I[K]$, the mutual information between $\mathbf{X}$ and $\mathbf{Y}$, under either the Joint Asymptotic Equipartition Theorem or under limitation by a distortion measure, through the Rate Distortion Theorem. The essential point is that $I[K]$ is a splitting criterion under these theorems, and thus partakes of the homology with free energy density which we have invoked above.

Activation of universality class tuning, the mean field model's version of attentional focusing, then becomes itself a punctuated event in response to increasing linkage between the organism and an external structured signal or some particular system of internal events.

This iterated argument exactly parallels the extension of the General Linear Model to the Hierarchical Linear Model in regression theory [15].

Another path to the fluctuating dynamic threshold might be through a second order iteration similar to that just above, but focused on the parameters defining the universality class distributions given above.

A network of dynamic manifolds and its tuning. The set of universality class tuning parameters, A_k, defines a manifold whose topology could also be more fully analyzed using Morse theory. That is an equivalence class of dynamic manifolds is determined, not by universality class, which is tunable, but by the underlying form of the renormalization relation, in the sense of the many different possible renormalization symmetries described above. Thus the possible higher level dynamic manifolds in this model are characterized by fixed renormalization relations, but tunable universality class parameters. One can then invoke a crosstalk coupling within a groupoid network of different dynamic manifolds defined by these renormalization relations, leading to the same kind of Morse theoretic analysis of the higher level topological structure.

Acknowledgments

The authors thank R.G. Wallace and M. Hodgkinson for useful suggestions, and the editors and reviewers for comments helpful in revision.

References

1. Adami, C., Cerf, N.: Physical complexity of symbolic sequences. Physica D 137, 62–69 (2000)
2. Adami, C., Ofria, C., Collier, T.: Evolution of biological complexity. Proceedings of the National Academy of Sciences 97, 4463–4468 (2000)
3. Albert, R., Barabasi, A.: Statistical mechanics of complex networks. Reviews of Modern Physics 74, 47–97 (2002)
4. Ancel, L.: A quantitative model of the Simpson-Baldwin effect. Journal of Theoretical Biology 196, 197–209 (1999)
5. Ash, R.: Information Theory. Dover Publications, New York (1990)
6. Atlan, H., Cohen, I.: Immune information, self-organization and meaning. International Immunology 10, 711–717 (1998)
7. Auslander, L.: Differential Geometry. Harper and Row, New York (1967)
8. Avital, E., Jablonka, E.: Animal Traditions: Behavioral inheritance in evolution. Cambridge University Press, New York (2000)
9. Baker, M., Stock, J.: Signal transduction: networks and integrated circuits in bacterial cognition. Current Biology 17(4), R1021–R1024 (2007)
10. Barkow, J., Cosmides, L., Tooby, J. (eds.): The Adapted Mind: Biological Approaches to Mind and Culture. University of Toronto Press (1992)
11. Beck, C., Schlogl, F.: Thermodynamics of Chaotic Systems. Cambridge University Press, Cambridge (1995)
12. Binney, J., Dowrick, N., Fisher, A., Newman, M.: The theory of critical phenomena. Clarendon Press, Oxford (1986)

13. Bonner, J.: The evolution of culture in animals. Princeton University Press, Princeton (1980)
14. Burago, D., Burago, Y., Ivanov, S.: A Course in Metric Geometry. American Mathematical Society, Providence (2001)
15. Byrk, A., Raudenbusch, S.: Hierarchical Linear Models: Applications and Data Analysis Methods. Sage Publications, New York (2001)
16. Champagnat, N., Ferriere, R., Meleard, S.: Unifying evolutionary dynamics: From individual stochastic processes to macroscopic models. Theoretical Population Biology 69, 297–321 (2006)
17. Cohen, I.: The cognitive principle challenges clonal selection. Immunology Today 13, 441–444 (1992)
18. Cohen, I.: Tending Adam's Garden: Evolving the Cognitive Immune Self. Academic Press, New York (2000)
19. Cohen, I.: Immune system computation and the immunological homunculus. In: Nierstrasz, O., Whittle, J., Harel, D., Reggio, G. (eds.) MoDELS 2006. LNCS, vol. 4199, pp. 499–512. Springer, Heidelberg (2006)
20. Cohen, I., Harel, D.: Explaining a complex living system: dynamics, multi-scaling and emergence. Journal of the Royal Society: Interface 4, 175–182 (2007)
21. Cover, T., Thomas, J.: Elements of Information Theory. John Wiley and Sons, New York (1991)
22. Dembo, A., Zeitouni, O.: Large Deviations: Techniques and Applications, 2nd edn. Springer, New York (1998)
23. Dercole, F., Ferriere, R., Gragnani, A., Rinaldi, S.: Coevolution of slow-fast populations: evolutionary sliding, evolutionary pseudo-equilibria and complex Red Queen dynamics. Proceedings of the Royal Society, B 273, 983–990 (2006)
24. Diekmann, U., Law, R.: The dynamical theory of coevolution: a derivation from stochastic ecological processes. Journal of Mathematical Biology 34, 579–612 (1996)
25. Dimitrov, A., Miller, J.: Neural coding and decoding: communication channels and quantization. Computation and Neural Systems 12, 441–472 (2001)
26. Dretske, F.: The explanatory role of information. Philosophical Transactions of the Royal Society A 349, 59–70 (1994)
27. Durham, W.: Coevolution: Genes, Culture, and Human Diversity. Stanford University Press, Palo Alto (1991)
28. Eldredge, N., Gould, S.: Punctuated equilibrium: an alternative to phyletic gradualism. In: Schopf, T. (ed.) Models in Paleobiology, pp. 82–115. Freeman, Cooper and Co., San Francisco (1972)
29. Eldredge, N.: Time Frames: The Rethinking of Darwinian Evolution and the Theory of Punctuated Equilibria. Simon and Schuster, New York (1985)
30. Emery, M.: Stochastic Calculus in Manifolds. Universitext series. Springer, New York (1989)
31. English, T.: Evaluation of evolutionary and genetic optimizers: no free lunch. In: Fogel, L., Angeline, P., Back, T. (eds.) Evolutionary Programming V: Proceedings of the Fifth Annual Conference on Evolutionary Programming, pp. 163–169. MIT Press, Cambridge (1996)
32. Fath, B., Cabezas, H., Pawlowski, C.: Regime changes in ecological systems: an information theory approach. Journal of Theoretical Biology 222, 517–530 (2003)
33. Feller, W.: An Introduction to Probability Theory and Its Applications. John Wiley and Sons, New York (1971)
34. Feynman, R.: Feynman Lectures on Computation. Addison-Wesley, Reading (1996)

35. Fleming, R., Shoemaker, C.: Evaluating models for spruce budworm-forest management: comparing output with regional field data. Ecological Applications 2, 460–477 (1992)
36. Forlenza, M., Baum, A.: Psychosocial influences on cancer progression: alternative cellular and molecular mechanisms. Current Opinion in Psychiatry 13, 639–645 (2000)
37. Franzosi, R., Pettini, M.: Theorem on the origin of phase transitions. Physical Review Letters 92, 060601 (2004)
38. Fredlin, M., Wentzell, A.: Random Perturbations of Dynamical Systems. Springer, New York (1998)
39. Glazebrook, J., Wallace, R.: Rate distortion manifolds as model spaces for cognitive information (submitted, 2007)
40. Goubault, E.: Some geometric perspectives in concurrency theory. Homology, Homotopy, and Applications 5, 95–136 (2003)
41. Goubault, E., Raussen, M.: Dihomotopy as a tool in state space analysis. In: Rajsbaum, S. (ed.) LATIN 2002. LNCS, vol. 2286, pp. 16–37. Springer, Heidelberg (2002)
42. Gould, S.: The Structure of Evolutionary Theory. Harvard University Press, Cambridge (2002)
43. Grossman, Z.: Round 3. Seminars in Immunology 12, 313–318 (2000)
44. Gunderson, L.: Ecological resilience – in theory and application. Annual Reviews of Ecological Systematics 31, 425–439 (2000)
45. Gunderson, L.: Personal communication (2007)
46. Hartl, D., Clark, A.: Principles of Population Genetics. Sinaur Associates, Sunderland (1997)
47. Holling, C.: Resilience and stability of ecological systems. Annual Reviews of Ecological Systematics 4, 1–23 (1973)
48. Holling, C.: Cross-scale morphology, geometry and dynamics of ecosystems. Ecological Monographs 41, 1–50 (1992)
49. Jimenez-Montano, M.: Formal languages and theoretical molecular biology. In: Goodwin, B., Saunders, P. (eds.) Theoretical Biology: Epigenetic an Evolutionary Order in Complex Systems. Edinburgh University Press (1989)
50. Kastner, M.: Phase transitions and configuration space topology (2006) (ArXiv preprint cond-mat/0703401)
51. Khinchin, A.: Mathematical Foundations of Information Theory. Dover Publications, New York (1957)
52. Laland, K., Odling-Smee, F., Feldman, M.: Evolutionary consequences of niche construction and their implications for ecology. Proceedings of the National Academy of Sciences 96, 10242–10247 (1999)
53. Lee, J.: Introduction to Topological Manifolds. Springer, New York (2000)
54. Levin, S.: Ecology in theory and application. In: Levin, S., Hallam, T., Gross, L. (eds.) Applied Mathematical Ecology biomathematics Texts, vol. 18. Springer, New York (1989)
55. Lewontin, R.: Biology as Ideology: The Doctrine of DNA. Harper Collins, New York (1993)
56. Lewontin, R.: The Triple Helix: gene, organism, and environment. Harvard University Press (2000)
57. Liao, J., Biscolo, R., Yang, Y., My Tran, L., Sabatti, C., Roychowdhury, V.: Network component analysis: Reconstruction of regulatory signals in biological systems. Proceedings of the National Academy of Sciences 100, 15522–15527 (2003)

58. Liu, Y., Ringner, M.: Revealing signaling pathway deregulation by using gene expression signatures and regulatory motif analysis. Genome Biology 8, R77 (2007)
59. Luchinsky, D.: On the nature of large fluctuations in equilibrium systems: observations of an optimal force. Journal of Physics A 30, L577–L583 (1997)
60. Matsumoto, Y.: An Introduction to Morse Theory, Translations of Mathematical Monographs, vol. 208. American Mathematical Society (2002)
61. Michel, L., Mozrymas, J.: Application of Morse Theory to the symmetry breaking in the Landau theory of the second order phase transition. In: Kramer, P., Rieckers, A. (eds.) Group Theoretical Methods in Physics: Sixth International Colloquium. Lecture Notes in Physics, vol. 79, pp. 447–461. Springer, New York (1977)
62. Milnor, J.: Morse Theory. Annals of Mathematical Studies, vol. 51. Princeton University Press, Princeton (1963)
63. Nunney, L.: Lineage selection and the evolution of multistage carcinogenesis. Proceedings of the London Royal Society B 266, 493–498 (1999)
64. Odling-Smee, F., Laland, K., Feldman, M.: Niche construction. The American Naturalist 147, 641–648 (1996)
65. Ofria, C., Adami, C., Collier, T.: Selective pressures on genomes in molecular evolution. Journal of Theoretical Biology 222, 477–483 (2003)
66. Onsager, L., Machlup, S.: Fluctuations and irreversible processes. Physical Review 91, 1505–1512 (1953)
67. O'Nuallain, S.: Code and context in gene expression, cognition, and consciousness. In: Barbieri, M., Hoffmeyer, J. (eds.) Biosemiotics: The codes of life. Springer, New York (2008)
68. Pettini, M.: Geometry and Topology in Hamiltonian Dynamics and Statistical Mechanics. Springer, New York (2007)
69. Pettini, M., Franzosi, R., Spinelli, L.: Topology and phase transitions I. Preliminary results. Nuclear Physics B 782, 189–218 (2007)
70. Pielou, E.C.: Mathematical Ecology. John Wiley and Sons, New York (1977)
71. Podolsky, S., Tauber, A.: The generation of diversity: Clonal selection theory and the rise of molecular biology. Harvard University Press (1998)
72. Pratt, V.: Modeling concurrency with geometry. In: Proceedings of the 18th ACM SIGPLAN-SIGACT Symposium on Principles of Programming Languages, pp. 311–322 (1991)
73. Priami, C.: Computational thinking in biology. In: Priami, C. (ed.) Transactions on Computational Systems Biology VIII. LNCS (LNBI), vol. 4780, pp. 63–76. Springer, Heidelberg (2007)
74. Ricotta, C.: Additive partition of parametric information and its associated β-diversity measure. Acta Biotheoretica 51, 91–100 (2003)
75. Ridley, M.: Evolution, 2nd edn. Blackwell Science/Oxford University Press (1996)
76. Rojdestvensky, I., Cottam, M.: Mapping of statistical physics to information theory with applications to biological systems. Journal of Theoretical Biology 202, 43–54 (2000)
77. Sayyed-Ahmad, A., Tuncay, K., Ortoleva, P.: Transcriptional regulatory network refinement and quantification through kinetic modeling, gene expression microarray data and information theory. BMC Bioinformatics 8, 20 (2007)
78. Shirkov, D., Kovalev, V.: The Bogoliubov renormalization group and solution symmetry in mathematical physics. Physics Reports 352, 219–249 (2001)
79. Soyer, O., Salathe, M., Bonhoeffer, S.: Signal transduction networks: Topology, response and biochemical processes. Journal of Theoretical Biology 238, 416–425 (2006)

80. Tauber, A.: Conceptual shifts in immunology: Comments on the 'two-way paradigm'. Theoretical Medicine and Bioethics 19, 457–473 (1998)
81. Volney, W., Fleming, R.: Spruce budworm *(Choristoneura spp.)* biotype reactions to forest and climate characteristics. Global Change Biology 13, 1630–1643 (2007)
82. Waddington, C.: Epilogue. In: Waddington, C. (ed.) Towards a Theoretical Biology: Essays. Aldine-Atherton, Chicago (1972)
83. Wallace, R., Wallace, R.G.: Psychopathica Automatorum: A cognitive neuroscience perspective on highly parallel computation and its dysfunctions (to appear, 2008)
84. Wallace, R., Fullilove, M.T.: Collective Consciousness and its Discontents: Institutional Distributed Cognition, Racial Policy and Public Health in the United States. Springer, New York (2008)
85. Wallace, R., Wallace, R.G.: Information theory, scaling laws and the thermodynamics of evolution. Journal of Theoretical Biology 192, 545–559 (1998)
86. Wallace, R., Wallace, R.G.: Organisms, organizations, and interactions: an information theory approach to biocultural evolution. BioSystems 51, 101–119 (1999)
87. Wallace, R., Wallace, R.G.: Immune cognition and vaccine strategy: beyond genomics. Microbes and Infection 4, 521–527 (2002)
88. Wallace, R., Wallace, D., Wallace, R.G.: Toward cultural oncology: the evolutionary information dynamics of cancer. Open Systems and Information Dynamics 10, 159–181 (2003)
89. Wallace, R.: Language and coherent neural amplification in hierarchical systems: renormalization and the dual information source of a generalized stochastic resonance. International Journal of Bifurcation and Chaos 10, 493–502 (2000)
90. Wallace, R.: Immune cognition and vaccine strategy: pathogenic challenge and ecological resilience. Open Systems and Information Dynamics 9, 51–83 (2002)
91. Wallace, R.: Adaptation, punctuation and rate distortion: non-cognitive 'learning plateaus' in evolutionary process. Acta Biotheoretica 50, 101–116 (2002)
92. Wallace, R.: Consciousness: A Mathematical Treatment of the Global Neuronal Workspace Model. Springer, New York (2005)
93. Wallace, R.: A global workspace perspective on mental disorders. Theoretical Biology and Medical Modelling 2, 49 (2005)
94. Whitham, T.: A framework for community and ecosystem genetics: from genes to ecosystems. Nature Reviews: genetics 7, 510–523 (2006)
95. Wilson, K.: Renormalization group and critical phenomena. I Renormalization group and the Kadanoff scaling picture. Physical Review B 4, 3174–3183 (1971)
96. Wymer, C.: Structural nonlinear continuous-time models in econometrics. Macroeconomic Dynamics 1, 518–548 (1997)
97. Zhu, R., Riberio, A., Salahub, D., Kauffman, S.: Studying genetic regulatory networks at the molecular level: Delayed reaction stochastic models. Journal of Theoretical Biology 246, 725–745 (2007)
98. Zurek, W.: Cosomological experiments in superfluid helium? Nature 317, 505–508 (1985)
99. Zurek, W.: Shards of broken symmetry. Nature 382, 296–298 (1996)

Stochastic Calculus of Looping Sequences for the Modelling and Simulation of Cellular Pathways

Roberto Barbuti[1], Andrea Maggiolo-Schettini[1], Paolo Milazzo[1], Paolo Tiberi[1], and Angelo Troina[2]

[1] Dipartimento di Informatica, Università di Pisa
Largo B. Pontecorvo 3, 56127 - Pisa, Italy
[2] Dipartimento di Informatica, Università di Torino
Corso Svizzera 185, 10149 - Torino Italy

Abstract. The paper presents the Stochastic Calculus of Looping Sequences (SCLS) suitable to describe microbiological systems, such as cellular pathways, and their evolution. Systems are represented by terms. The terms of the calculus are constructed by basic constituent elements and operators of sequencing, looping, containment and parallel composition. The looping operator allows tying up the ends of a sequence, thus creating a circular sequence which can represent a membrane.

The evolution of a term is modelled by a set of rewrite rules enriched with stochastic rates representing the speed of the activities described by the rules, and can be simulated automatically.

As applications, we give SCLS representations of the regulation process of the lactose operon in *Escherichia coli* and of the quorum sensing in *Pseudomonas aeruginosa*.

A prototype simulator (SCLSm) has been implemented in F# and used to run the experiments. A public version of the tool is available at the url: `http://www.di.unipi.it/~milazzo/biosims/`.

1 Introduction

Biologists usually describe biological systems by mathematical means, such as differential equations. This allows them to reason on the behaviour of the described systems and to perform simulations. Mathematical modelling becomes more difficult both in specification and in analysis when the complexity of the system increases. This is one of the main motivations for the application of Computer Science formalisms to the description of biological systems [27]. Another motivation is that the use of formal means of Computer Science permits the application of analysis methods that are practically unknown to biologists, such as model checking.

Among the formalisms that either have been applied to or have been inspired by biological systems there are automata-based models [1,19], rewrite systems [12,21], and process calculi [27,25,7]. Automata have the advantage of allowing the direct use of many verification tools such as model checkers. Rewrite systems usually allow describing biological systems with a notation that can be easily

C. Priami (Ed.): Trans. on Comput. Syst. Biol. IX, LNBI 5121, pp. 86–113, 2008.

understood by biologists. On the other hand, automata-like models and rewrite systems present, in general, problems from the point of view of compositionality. Compositionality allows studying the behaviour of a system componentwise, and is in general ensured by process calculi, included those commonly used to describe biological systems.

In [4,5,20] we developed a new formalism, called Calculus of Looping Sequences (CLS for short), for describing biological systems and their evolution. CLS is based on term rewriting with some features, such as a commutative parallel composition operator, and some semantic means, such as bisimulations [5,6], which are common in process calculi. This permits to combine the simplicity of notation of rewrite systems with the advantage of a form of compositionality.

In this paper we focus on quantitative aspects of our formalism, in particular, to model speed of activities, we develop a stochastic extension of CLS (called SCLS). Rates are associated with rewrite rules in order to model the speed of the described activities. Therefore, transitions derived in SCLS are driven by a rate that models the parameter of an exponential distribution and characterizes the stochastic behaviour of the transition. The choice of the next rule to be applied and of the time of its application is based on the classical Gillespie's algorithm [14].

We have developed a prototype simulator for SCLS. To show the expressiveness of our formalism, we model and simulate two examples: the regulation of the lactose operon in *Escherichia coli* and the quorum sensing in *Pseudomonas aeruginosa*. The first example shows all the features of SCLS used to describe a classical model. The second one shows the merit of the computational approach with respect to mathematical modelling when the complexity of the system increases.

We remark that the contribution of this paper is not in the simulation algorithm, inspired by the standard Gillespie's algorithm, but in the language proposed to describe systems: it allows describing cellular structures and compartments, and this simplifies the modelling of a cell as a system whose components are described individually.

1.1 Summary

The remainder of this paper is organized as follows. In Section 2 we formally recall the Calculus of Looping Sequence and we give some guidelines for the modelling of biological systems. In Section 3 we introduce our stochastic extension. In Sections 4 and 5 we use the stochastic framework to model and analyse two different applications; namely, we model the lactose operon of *Escherichia Coli* and a quorum sensing process in *Pseudomonas aeruginosa*. Finally, in Section 6 we draw our conclusions and we present some related work.

2 The Calculus of Looping Sequences

In this section we recall the Calculus of Looping Sequences (CLS). It is based on term rewriting, and hence a CLS model consists of a term and a set of rewrite rules. The term represents the structure of the modelled system, and the rewrite rules represent its evolution.

We start with defining the syntax of terms. We assume a possibly infinite alphabet $\mathcal{E}$ of symbols ranged over by $a, b, c, \ldots$.

Definition 1 (Terms). Terms T *and* Sequences S *of CLS are given by the grammars:*

$$T ::= S \mid (S)^L \rfloor T \mid T \mid T$$
$$S ::= \epsilon \mid a \mid S \cdot S$$

where a is any element of $\mathcal{E}$ and ϵ is the empty sequence. We denote the infinite sets of terms and sequences with $\mathcal{T}$ and $\mathcal{S}$, respectively.

In CLS we have a sequencing operator $\_ \cdot \_$, a looping operator $(\_)^L$, a parallel composition operator $\_ \mid \_$, and a containment operator $\_ \rfloor \_$. Sequencing can be used to concatenate elements of the alphabet $\mathcal{E}$. The empty sequence ϵ denotes the concatenation of zero symbols. A term can be either a sequence, or a looping sequence (that is the application of the looping operator to a sequence) containing another term, or the parallel composition of two terms. By the definition of terms, we have that looping and containment are always applied together, hence we can consider them as a single binary operator $(\_)^L \rfloor \_$ that applies to one sequence and one term.

The biological interpretation of the operators is the following: the main entities which occur in cells are DNA and RNA strands, proteins, membranes, and other macro-molecules. DNA strands (and similarly RNA strands) are sequences of nucleic acids, but they can be seen also at a higher level of abstraction as sequences of genes. Proteins are sequences of amino acids which usually have a very complex three-dimensional structure. In a protein there are usually (relatively) few subsequences, called domains, which actually are able to interact with other entities by means of chemical reactions. CLS sequences can model DNA/RNA strands and proteins by describing each gene or each domain with a symbol of the alphabet. Membranes are closed surfaces often interspersed with proteins, and may have a content. A closed surface can be modelled by a looping sequence. The elements (or the subsequences) of the looping sequence may represent the proteins on the membrane, and by the containment operator it is possible to specify what the membrane contains. Other macro-molecules can be modelled as single alphabet symbols, or as sequences of their components. Finally, juxtaposition of entities can be described by the parallel composition operator of their representations. A deeper description of the biological interpretation of CLS operators together with some modelling guidelines will be given in Section 2.1.

Brackets can be used to indicate the order of application of the operators, and $(\_)^L \rfloor \_$ has the precedence over $\_ \mid \_$. An example of CLS term is $a \mid b \mid (m \cdot n)^L \rfloor (c \cdot d \mid e)$. It represents a membrane with two molecules m and n (for instance, two proteins) on its surface, and containing a sequence $c \cdot d$ (for instance, a DNA

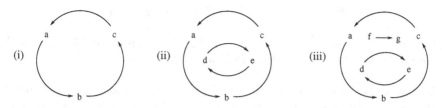

Fig. 1. (i) represents $(a \cdot b \cdot c)^L \rfloor \epsilon$; (ii) represents $(a \cdot b \cdot c)^L \rfloor (d \cdot e)^L \rfloor \epsilon$; (iii) represents $(a \cdot b \cdot c)^L \rfloor (((d \cdot e)^L \rfloor \epsilon) \mid f \cdot g)$

strand) and a molecule e. Molecules a and b are outside the membrane. See Figure 1 for some graphical representations.

In CLS we may have syntactically different terms representing the same structure. We introduce structural congruence relations to identify such terms.

Definition 2 (Structural Congruence). *The structural congruence relations* $\equiv_S$ *and* $\equiv_T$ *are the least congruence relations on sequences and on terms, respectively, satisfying the following rules:*

$$S_1 \cdot (S_2 \cdot S_3) \equiv_S (S_1 \cdot S_2) \cdot S_3 \qquad S \cdot \epsilon \equiv_S \epsilon \cdot S \equiv_S S$$

$$S_1 \equiv_S S_2 \text{ implies } S_1 \equiv_T S_2 \text{ and } (S_1)^L \rfloor T \equiv_T (S_2)^L \rfloor T$$

$$T_1 \mid T_2 \equiv_T T_2 \mid T_1 \qquad T_1 \mid (T_2 \mid T_3) \equiv_T (T_1 \mid T_2) \mid T_3 \qquad T \mid \epsilon \equiv_T T$$

$$(\epsilon)^L \rfloor \epsilon \equiv_T \epsilon \qquad (S_1 \cdot S_2)^L \rfloor T \equiv_T (S_2 \cdot S_1)^L \rfloor T$$

Rules of the structural congruence state the associativity of $\cdot$ and $\mid$, the commutativity of the latter and the neutral role of ϵ. Moreover, axiom $(S_1 \cdot S_2)^L \rfloor T \equiv_T (S_2 \cdot S_1)^L \rfloor T$ says that looping sequences can rotate. In the following we will use $\equiv$ in place of $\equiv_T$.

Rewrite rules are defined essentially as pairs of terms, in which the first term describes the portion of the system in which the event modelled by the rule may occur, and the second term describes how that portion of the system changes when the event occurs. In the terms of a rewrite rule we allow the use of variables. As a consequence, a rule will be applicable to all terms which can be obtained by properly instantiating its variables. Variables can be of three kinds: two are associated with the two different syntactic categories of terms and sequences, and one is associated with single alphabet elements. We assume a set of term variables TV ranged over by $X, Y, Z, \ldots$, a set of sequence variables SV ranged over by $\tilde{x}, \tilde{y}, \tilde{z}, \ldots$, and a set of element variables $\mathcal{X}$ ranged over by $x, y, z, \ldots$. All these sets are pairwise disjoint and possibly infinite. We denote by $\mathcal{V}$ the set of all variables $TV \cup SV \cup \mathcal{X}$, and with ρ any variable in $\mathcal{V}$. A pattern is a term which may include variables.

Definition 3 (Patterns). *Patterns P and sequence patterns SP of CLS are given by the following grammar:*

$$P ::= SP \mid (SP)^L \rfloor P \mid P \mid P \mid X$$

$$SP \quad ::= \epsilon \quad | \quad a \quad | \quad SP \cdot SP \quad | \quad \tilde{x} \quad | \quad x$$

where a is an element of $\mathcal{E}$, and $X, \tilde{x}$ and x are elements of TV, SV and $\mathcal{X}$, respectively. We denote with $\mathcal{P}$ the infinite set of patterns.

We assume the structural congruence relation to be trivially extended to patterns. An *instantiation* is a partial function $\sigma : \mathcal{V} \to \mathcal{T}$. An instantiation must preserve the type of variables, thus for $X \in TV, \tilde{x} \in SV$ and $x \in \mathcal{X}$ we have $\sigma(X) \in \mathcal{T}, \sigma(\tilde{x}) \in \mathcal{S}$, and $\sigma(x) \in \mathcal{E}$, respectively. Given $P \in \mathcal{P}$, with $P\sigma$ we denote the term obtained by replacing each occurrence of each variable $\rho \in \mathcal{V}$ appearing in P with the corresponding term $\sigma(\rho)$. With Σ we denote the set of all the possible instantiations, and, given $P \in \mathcal{P}$, with $Var(P)$ we denote the set of variables appearing in P. Now we can define rewrite rules.

Definition 4 (Rewrite Rules). *A rewrite rule is a pair of patterns (P_1, P_2), denoted with $P_1 \mapsto P_2$, where $P_1, P_2 \in \mathcal{P}$, $P_1 \not\equiv \epsilon$ and such that $Var(P_2) \subseteq Var(P_1)$.*

A rewrite rule $P_1 \mapsto P_2$ states that a term $P_1\sigma$, obtained by instantiating variables in P_1 by some instantiation function σ, can be transformed into the term $P_2\sigma$. We define the semantics of CLS as a transition system, in which states correspond to terms, and transitions correspond to rule applications. The semantics of CLS is defined by resorting to the notion of contexts.

Definition 5 (Contexts). *Contexts C are defined as:*

$$C ::= \square \quad | \quad C \,|\, T \quad | \quad T \,|\, C \quad | \quad (S)^L \rfloor C$$

where $T \in \mathcal{T}$ and $S \in \mathcal{S}$. The context $\square$ is called the empty *context. We denote with $\mathcal{C}$ the infinite set of contexts.*

By definition, every context contains a single $\square$. Let us assume $C, C' \in \mathcal{C}$. With $C[T]$ we denote the term obtained by replacing $\square$ with T in C; with $C[C']$ we denote context composition, whose result is the context obtained by replacing $\square$ with C' in C. The structural congruence relation can be easily extended to contexts, namely $C \equiv C'$ if and only if $C[\epsilon] \equiv C'[\epsilon]$.

Rewrite rules can be applied to terms only if they occur in a legal context. Note that the general form of rewrite rules does not permit to have sequences as contexts. A rewrite rule introducing a parallel composition on the right hand side (as $a \mapsto b \,|\, c$) applied to an element of a sequence (e.g., $m \cdot a \cdot m$) would result into a syntactically incorrect term (in this case $m \cdot (b \,|\, c) \cdot m$). To modify a sequence, a pattern representing the whole sequence must appear in the rule. For example, rule $a \cdot \tilde{x} \mapsto a \,|\, \tilde{x}$ can be applied to any sequence starting with element a, and, hence, the term $a \cdot b$ can be rewritten as $a \,|\, b$, and the term $a \cdot b \cdot c$ can be rewritten as $a \,|\, b \cdot c$.

The semantics of CLS is defined as follows.

Table 1. Guidelines for the abstraction of biomolecular entities into CLS

Biomolecular Entity	CLS Term
Elementary object (genes, domains, other molecules, etc...)	Alphabet symbol
DNA strand	Sequence of elements repr. genes
RNA strand	Sequence of elements repr. transcribed genes
Protein	Sequence of elements repr. domains or single alphabet symbol
Molecular population	Parallel composition of molecules
Membrane	Looping sequence

Definition 6 (Semantics). *Given a finite set of rewrite rules $\mathcal{R}$, the semantics of CLS is the least relation closed with respect to $\equiv$ and satisfying the following inference rule:*

$$\frac{P_1 \mapsto P_2 \in \mathcal{R} \qquad P_1\sigma \not\equiv \epsilon \qquad \sigma \in \Sigma \qquad C \in \mathcal{C}}{C[P_1\sigma] \to C[P_2\sigma]}$$

2.1 Modelling Guidelines

We describe how CLS can be used to model biomolecular systems analogously to what done by Regev and Shapiro in [28] for the π-calculus. An abstraction is a mapping from a real-world domain to a mathematical domain, which may allow highlighting some essential properties of a system while ignoring other, complicating, ones. In [28], Regev and Shapiro show how to abstract biomolecular systems as concurrent computations by identifying the biomolecular entities and events of interest and by associating them with concepts of concurrent computations such as concurrent processes and communications. In particular, they give some guidelines for the abstraction of biomolecular systems to the π-calculus, and give some simple examples.

The use of rewrite systems, such as CLS, to describe biological systems is founded on a different abstraction. Usually, entities (and their structures) are abstracted by terms of the rewrite system, and events by rewrite rules. We have already introduced the biological interpretation of CLS operators in the previous section. Here we want to give more general guidelines.

First of all, we should select the biomolecular entities of interest. Since we want to describe cells, we consider molecular populations and membranes. Molecular populations are groups of molecules that are in the same compartment of the cell. As we have said before, molecules can be of many types: we classify them as DNA and RNA strands, proteins, and other molecules. Membranes are considered as elementary objects, in the sense that we do not describe them at the level of the lipids they are made of. The only interesting properties of a membrane are that it may have a content (hence, create a compartment) and that it may have molecules on its surface.

Table 2. Guidelines for the abstraction of biomolecular events into CLS

Biomolecular Event	Examples of CLS Rewrite Rule
State change	$a \mapsto b$ $\widetilde{x} \cdot a \cdot \widetilde{y} \mapsto \widetilde{x} \cdot b \cdot \widetilde{y}$
Complexation	$a \,\vert\, b \mapsto c$ $\widetilde{x} \cdot a \cdot \widetilde{y} \,\vert\, b \mapsto \widetilde{x} \cdot c \cdot \widetilde{y}$
Decomplexation	$c \mapsto a \,\vert\, b$ $\widetilde{x} \cdot c \cdot \widetilde{y} \mapsto \widetilde{x} \cdot a \cdot \widetilde{y} \,\vert\, b$
Catalysis	$c \,\vert\, P_1 \mapsto c \,\vert\, P_2$ where $P_1 \mapsto P_2$ is the catalyzed event
State change on membrane	$(a \cdot \widetilde{x})^L \rfloor X \mapsto (b \cdot \widetilde{x})^L \rfloor X$
Complexation on membrane	$(a \cdot \widetilde{x} \cdot b \cdot \widetilde{y})^L \rfloor X \mapsto (c \cdot \widetilde{x} \cdot \widetilde{y})^L \rfloor X$ $a \,\vert\, (b \cdot \widetilde{x})^L \rfloor X \mapsto (c \cdot \widetilde{x})^L \rfloor X$ $(b \cdot \widetilde{x})^L \rfloor (a \,\vert\, X) \mapsto (c \cdot \widetilde{x})^L \rfloor X$
Decomplexation on membrane	$(c \cdot \widetilde{x})^L \rfloor X \mapsto (a \cdot b \cdot \widetilde{x})^L \rfloor X$ $(c \cdot \widetilde{x})^L \rfloor X \mapsto a \,\vert\, (b \cdot \widetilde{x})^L \rfloor X$ $(c \cdot \widetilde{x})^L \rfloor X \mapsto (b \cdot \widetilde{x})^L \rfloor (a \,\vert\, X)$
Catalysis on membrane	$(c \cdot \widetilde{x} \cdot SP_1 \cdot \widetilde{y})^L \rfloor X \mapsto (c \cdot \widetilde{x} \cdot SP_2 \cdot \widetilde{y})^L \rfloor X$ where $SP_1 \mapsto SP_2$ is the catalyzed event
Membrane crossing	$a \,\vert\, (\widetilde{x})^L \rfloor X \mapsto (\widetilde{x})^L \rfloor (a \,\vert\, X)$ $(\widetilde{x})^L \rfloor (a \,\vert\, X) \mapsto a \,\vert\, (\widetilde{x})^L \rfloor X$ $\widetilde{x} \cdot a \cdot \widetilde{y} \,\vert\, (\widetilde{z})^L \rfloor X \mapsto (\widetilde{z})^L \rfloor (\widetilde{x} \cdot a \cdot \widetilde{y} \,\vert\, X)$ $(\widetilde{z})^L \rfloor (\widetilde{x} \cdot a \cdot \widetilde{y} \,\vert\, X) \mapsto \widetilde{x} \cdot a \cdot \widetilde{y} \,\vert\, (\widetilde{z})^L \rfloor X$
Catalyzed membrane crossing	$a \,\vert\, (b \cdot \widetilde{x})^L \rfloor X \mapsto (b \cdot \widetilde{x})^L \rfloor (a \,\vert\, X)$ $(b \cdot \widetilde{x})^L \rfloor (a \,\vert\, X) \mapsto a \,\vert\, (b \cdot \widetilde{x})^L \rfloor X$ $\widetilde{x} \cdot a \cdot \widetilde{y} \,\vert\, (b \cdot \widetilde{z})^L \rfloor X \mapsto (b \cdot \widetilde{z})^L \rfloor (\widetilde{x} \cdot a \cdot \widetilde{y} \,\vert\, X)$ $(b \cdot \widetilde{z})^L \rfloor (\widetilde{x} \cdot a \cdot \widetilde{y} \,\vert\, X) \mapsto \widetilde{x} \cdot a \cdot \widetilde{y} \,\vert\, (b \cdot \widetilde{z})^L \rfloor X$
Membrane joining	$(\widetilde{x})^L \rfloor (a \,\vert\, X) \mapsto (a \cdot \widetilde{x})^L \rfloor X$ $(\widetilde{x})^L \rfloor (\widetilde{y} \cdot a \cdot \widetilde{z} \,\vert\, X) \mapsto (\widetilde{y} \cdot a \cdot \widetilde{z} \cdot \widetilde{x})^L \rfloor X$
Catalyzed membrane joining	$(b \cdot \widetilde{x})^L \rfloor (a \,\vert\, X) \mapsto (a \cdot b \cdot \widetilde{x})^L \rfloor X$ $(\widetilde{x})^L \rfloor (a \,\vert\, b \,\vert\, X) \mapsto (a \cdot \widetilde{x})^L \rfloor (b \,\vert\, X)$ $(b \cdot \widetilde{x})^L \rfloor (\widetilde{y} \cdot a \cdot \widetilde{z} \,\vert\, X) \mapsto (\widetilde{y} \cdot a \cdot \widetilde{z} \cdot \widetilde{x})^L \rfloor X$ $(\widetilde{x})^L \rfloor (\widetilde{y} \cdot a \cdot \widetilde{z} \,\vert\, b \,\vert\, X) \mapsto (\widetilde{y} \cdot a \cdot \widetilde{z} \cdot \widetilde{x})^L \rfloor (b \,\vert\, X)$
Membrane fusion	$(\widetilde{x})^L \rfloor (X) \,\vert\, (\widetilde{y})^L \rfloor (Y) \mapsto (\widetilde{x} \cdot \widetilde{y})^L \rfloor (X \,\vert\, Y)$
Catalyzed membrane fusion	$(a \cdot \widetilde{x})^L \rfloor (X) \,\vert\, (b \cdot \widetilde{y})^L \rfloor (Y) \mapsto$ $\qquad (a \cdot \widetilde{x} \cdot b \cdot \widetilde{y})^L \rfloor (X \,\vert\, Y)$
Membrane division	$(\widetilde{x} \cdot \widetilde{y})^L \rfloor (X \,\vert\, Y) \mapsto (\widetilde{x})^L \rfloor (X) \,\vert\, (\widetilde{y})^L \rfloor (Y)$
Catalyzed membrane division	$(a \cdot \widetilde{x} \cdot b \cdot \widetilde{y})^L \rfloor (X \,\vert\, Y) \mapsto$ $\qquad (a \cdot \widetilde{x})^L \rfloor (X) \,\vert\, (b \cdot \widetilde{y})^L \rfloor (Y)$

Now, we select the biomolecular events of interest. The simplest kind of event is the change of state of an elementary object. Then, we may have interactions between molecules: in particular complexation, decomplexation and catalysis. These interactions may involve single elements of non-elementary molecules

(DNA and RNA strands, and proteins). Moreover, we may have interactions between membranes and molecules: in particular a molecule may cross or join a membrane. Finally, we may have interactions between membranes: in this case there may be many kinds of interactions (fusion, division, etc. . .).

The guidelines for the abstraction of biomolecular entities and events into CLS are given in Table 1 and Table 2, respectively. Entities are associated with CLS terms: elementary objects are modelled as alphabet symbols, non-elementary objects as CLS sequences and membranes as looping sequences. Biomolecular events are associated with CLS rewrite rules. In the table we give some examples of rewrite rules for each type of event. The list of examples is not complete: one could imagine also rewrite rules for the description of complexation/decomplexation events involving more than two molecules, or catalysis events in which the catalyzing molecule is on a membrane and the catalyzed event occurs in its content, or more complex interactions between membranes. We remark that in the second example of rewrite rule associated with the complexation event we have that one of the two molecules which are involved should be either an elementary object or a protein modelled as a single alphabet symbol. As before, this is caused by the problem of modelling protein interaction at the domain level. This problem is solved by an extension of CLS where links are considered. Such a model, called LCLS, is formalized in [3].

3   The Stochastic Calculus of Looping Sequences

The standard way of extending a formalism to model quantitative aspects of biological systems is by incorporating a collision-based stochastic framework on the lines of the one presented by Gillespie in [14]. Following the law of mass action, we need to count the number of reactants that are present in a system in order to compute the exact rate of a reaction. This has been done, for instance, for the π-calculus [23,25]. The idea of Gillespie's algorithm is that a rate constant is associated with each considered chemical reaction. Such a constant is obtained by multiplying the kinetic constant of the reaction by the number of possible combinations of reactants that may occur in the system. The resulting rate is then used as the parameter of an exponential distribution modelling the time spent between two occurrences of the considered chemical reaction.

The use of exponential distributions to represent the (stochastic) time spent between two occurrences of chemical reactions allows describing the system as a Continuous Time Markov Chain (CTMC), and consequently allows verifying properties of the described system analytically and by means of stochastic model checkers.

We start by adding rates to rewrite rules.

Definition 7 (Stochastic Rewrite Rule). *A stochastic rewrite rule is a triple* (P_1, P_2, k), *denoted with* $P_1 \overset{k}{\mapsto} P_2$, *where* $P_1, P_2 \in \mathcal{P}$, $P_1 \not\equiv \epsilon$ *and such that* $Var(P_2) \subseteq Var(P_1)$; $k \in \mathbb{R}^{\geq 0}$ *is the* rewrite rate.

To describe the evolution of a term, the stochastic semantics must consider, besides the rate of a rule, also the number of occurrences of subterms to which the rule can be applied and the terms produced. Subterms to which the rule can be applied correspond to reactants in a biological system. In what follows, a *subterm* of a term T will be a term $T' \not\equiv \epsilon$ for which a context C exists such that $T \equiv C[T']$, and a *reactant* will be an occurrence in T of a subterm.

Example 1. If $T = a \,|\, a \,|\, b \,|\, b$, then the set of subterms of T is

$$\{a \,,\, b \,,\, a\,|\,a \,,\, a\,|\,b \,,\, b\,|\,b \,,\, a\,|\,a\,|\,b \,,\, a\,|\,b\,|\,b \,,\, T\}$$

while

$$\{a \,,\, a \,,\, b \,,\, b \,,\, a\,|\,a \,,\, a\,|\,b \,,\, a\,|\,b \,,\, a\,|\,b \,,\, a\,|\,b \,,\, b\,|\,b \,,\, a\,|\,a\,|\,b \,,\, a\,|\,a\,|\,b \,,\, a\,|\,b\,|\,b \,,\, a\,|\,b\,|\,b \,,\, T\}$$

is the multiset of reactants in T. $\qquad\qquad\qquad\qquad\qquad\qquad\qquad\qquad\qquad\square$

Now, defining the stochastic semantics would be easy if rules would contain no variables. For instance, if we have the rewrite rule $a \,|\, b \stackrel{k}{\mapsto} c$, where k is the kinetic constant of the modelled chemical reaction, then its application rate is k multiplied by the number of possible combinations of occurrences of a and b in the term, namely the number of occurrences of $a \,|\, b$ in the multiset of reactants of the term. For example, given the term T in Example 1, we have two occurrences of a and two of b, hence the number of possible combinations of reactants is $2 \times 2 = 4$, and this holds also in the multiset of reactants of T, which contains four instances of $a \,|\, b$.

As we have variables, we have to take into account how they can be instantiated in order to compute the application rate of the rewrite rule. Variables allow a rewrite rule to stand for a family of ground rules, which represents a family of chemical reactions. Moreover, it often happens that the application rate of a rewrite rule depends on how many molecules of some kind are contained in the part of the system represented by a variable. For instance, consider a rule such as $a \,|\, (b\cdot\widetilde{x})^L \,\rfloor\, X \stackrel{k}{\mapsto} (c\cdot\widetilde{x})^L \,\rfloor\, X$, representing the binding of molecule a with an instance of b (resulting into the product molecule c) placed on the membrane represented by the looping sequence. We should have that the application rate of the derived reactions is proportional to the number of b which are present on the membrane, that is the number of b in the instantiation of the variable $\widetilde{x}$ plus one.

We remark that this problem has not been faced during the development of the stochastic extension of other formalisms such as the π-calculus, as those formalisms are not able to model chemical reactions with variables (as CLS patterns are). Also Gillespie's work does not deal with variables in the simulated chemical reactions. As a consequence, we have to give a reasonable interpretation to rewrite rules with variables.

We follow an approach on the lines of the one used by Krivine et al. in [17] for defining a stochastic semantics for Bigraphical Reactive Systems. The technique they have developed to count the occurrences of a reactant is based on the definition of *abstract* and *concrete* bigraphs. Here we consider as *abstract* the CLS terms and patterns defined as in Definitions 1 and 3. In the remainder of this section, we denote abstract terms and patterns using a tilde, as in $\tilde{T}$ and $\tilde{P}$. Now, we give the definition of *concrete* CLS patterns. Since a term is a ground pattern, the analogous definition for concrete terms can be inherited from the one for patterns.

Definition 8 (Concrete patterns and terms). *If $\tilde{P}$ is an abstract pattern, then a* concrete *pattern P, called a* concretion *of $\tilde{P}$, is obtained by assigning to each alphabet symbol syntactically appearing in $\tilde{P}$ a unique identifier $v \in Id$, where Id is a finite set of identifiers. With $`\mathcal{P}$ and $`\mathcal{T}$ we denote the sets of concrete patterns and terms, respectively.*

Intuitively, each symbol of the alphabet $\mathcal{E}$ appearing in patterns and terms, becomes unique in the concretion by labelling it with a fresh identifier. Moreover, we equip concrete patterns and terms with a notion of *support*.

Definition 9 (Support). *Given a concrete pattern P, we call* support *the set of identifiers used to label its alphabet symbols and we denote it with $Supp(P)$.*

Two concrete patterns P and P' are support-equivalent, *written $P \simeq P'$, if they differ only by a bijection between their supports which preserves structure. Namely, $P \simeq P'$ if and only if $\tilde{P} \equiv \tilde{P}'$ and there exists a bijection between $Supp(P)$ and $Supp(P')$. We denote the $\simeq$-equivalence class of P by $[P]$.*

As before, the analogous definitions of support and support-equivalence for concrete terms is inherited.

Given an abstract pattern $\tilde{P} = a \cdot \tilde{x} \mid (a \cdot b)^L \rfloor X \in \mathcal{P}$ concretions of $\tilde{P}$ are $P_1 = a_{v_1} \cdot \tilde{x} \mid (a_{v_2} \cdot b_{v_3})^L \rfloor X$ or $P_2 = a_{u_1} \cdot \tilde{x} \mid (a_{u_2} \cdot b_{u_3})^L \rfloor X$ with supports $Supp(P_1) = \{v_1, v_2, v_3\}$ and $Supp(P_2) = \{u_1, u_2, u_3\}$. Note that for an abstract pattern $\tilde{P}$ and a set of identifiers Id there exist many different concretions. For the case above we have, however, $P_1 \simeq P_2$.

We can extend the definition of concrete patterns to stochastic rewrite rules.

Definition 10. *If $R = (\tilde{P}_1, \tilde{P}_2, k)$ is a stochastic rewrite rule, then (P_1, P_2, k) is called a* concretion *of R.*

The definition of contexts is extended to deal with concrete terms in the natural way, with $`\mathcal{C}$ we denote the set of concrete contexts. Without loss of generality, we assume instantiations to return abstract or concrete terms when applied to abstract or concrete patterns respectively. Namely, given $\tilde{P} \in \mathcal{P}$, $\tilde{P}\sigma \in \mathcal{T}$, while given $P \in `\mathcal{P}$, $P\sigma \in `\mathcal{T}$.

Since we would like to define rewrite rules in an abstract way, we should define a notion of occurrence of abstract patterns within a term.

Definition 11 (Occurrences). *If $\tilde{P} \in \mathcal{P}$ is an abstract pattern and $T \in `\mathcal{T}$ a concrete term, an* occurrence *of $\tilde{P}$ in $T$ is a pair $(C, P)$, where $P \in `\mathcal{P}$ is a concretion of $\tilde{P}$ and $C \in `\mathcal{C}$ is a context such that $T \equiv C[P\sigma]$ for some instantiation σ.*

An occurrence *of a rule $R = (\tilde{P}_1, \tilde{P}_2, k)$ in a concrete term T is a pair (C, P_1), where (P_1, P_2, k) is a concretion of R and $T \equiv C[P_1\sigma]$ for some instantiation σ.*

If also $T' \simeq C[P_2\sigma]$ we say that the occurrence of rule R in T results into a term support-equivalent to T'. With $\mathcal{O}(R, T, \llbracket T' \rrbracket)$ we define the set of occurrences of rule R in the term T resulting in a concrete term support-equivalent to T'.

Example 2. Consider a concretion $T = a_{v_1} \mid a_{v_2} \mid b_{v_3} \mid b_{v_4}$ of the term in Example 1 and the abstract stochastic rewrite rule $R = a \mid b \overset{k}{\mapsto} c$. The occurrences of R in T are:

- $(a_{v_1} \mid b_{v_3} \mid \square \,,\ a_{v_2} \mid b_{v_4})$;
- $(a_{v_1} \mid b_{v_4} \mid \square \,,\ a_{v_2} \mid b_{v_3})$;
- $(a_{v_2} \mid b_{v_3} \mid \square \,,\ a_{v_1} \mid b_{v_4})$;
- $(a_{v_2} \mid b_{v_4} \mid \square \,,\ a_{v_1} \mid b_{v_3})$.

Moreover, all these occurrences result into a concrete term which is support-equivalent to $T' = a_{v_1} \mid c_{t_1} \mid b_{v_3}$. Thus, $\mathcal{O}(R, T, \llbracket T' \rrbracket)$ contains exactly the four occurrences listed above.

If we consider a term $T_o = a_{u_0} \mid a_{u_1} \mid (b_{u_2} \cdot c_{u_3} \cdot b_{u_4} \cdot a_{u_5})^L \rfloor \epsilon$ and the abstract rule $R' = a \mid (b \cdot \widetilde{x})^L \rfloor X \overset{k'}{\mapsto} (c \cdot \widetilde{x})^L \rfloor X$ (which contains variables), then the occurrences of R' in T_o are:

- $o_1 = (a_{u_1} \mid \square \,,\ a_{u_0} \mid (b_{u_2} \cdot \widetilde{x})^L \rfloor X)$, for $\sigma(\widetilde{x}) = c_{u_3} \cdot b_{u_4} \cdot a_{u_5}$ and $\sigma(X) = \epsilon$;
- $o_2 = (a_{u_1} \mid \square \,,\ a_{u_0} \mid (b_{u_4} \cdot \widetilde{x})^L \rfloor X)$, for $\sigma(\widetilde{x}) = a_{u_5} \cdot b_{u_2} \cdot c_{u_3}$ and $\sigma(X) = \epsilon$;
- $o_3 = (a_{u_0} \mid \square \,,\ a_{u_1} \mid (b_{u_2} \cdot \widetilde{x})^L \rfloor X)$, for $\sigma(\widetilde{x}) = c_{u_3} \cdot b_{u_4} \cdot a_{u_5}$ and $\sigma(X) = \epsilon$;
- $o_4 = (a_{u_0} \mid \square \,,\ a_{u_1} \mid (b_{u_4} \cdot \widetilde{x})^L \rfloor X)$, for $\sigma(\widetilde{x}) = a_{u_5} \cdot b_{u_2} \cdot c_{u_3}$ and $\sigma(X) = \epsilon$.

Note that these occurrences take into account all the possible combinations of any molecule a (outside the looping sequence) with any molecule b (in the looping sequence).

In this case, the different occurrences of the rule produce terms which are also structurally different. For example, by applying the first occurrence with the concretion of the right hand side of R' given by $P_2 = (c_{t_1} \cdot \widetilde{x})^L \rfloor X$, we get $T'_{o_1} = a_{u_1} \mid (c_{t_1} \cdot c_{u_3} \cdot b_{u_4} \cdot a_{u_5})^L \rfloor \epsilon$. Differently, if we apply the second occurrence, again with the same concretion $P_2 = (c_{t_1} \cdot \widetilde{x})^L \rfloor X$, we get $T'_{o_2} = a_{u_1} \mid (c_{t_1} \cdot a_{u_5} \cdot b_{u_2} \cdot c_{u_3})^L \rfloor \epsilon$. Note that T'_{o_1} and T'_{o_2} are structurally different: in T'_{o_1}, the a-molecule remaining in the looping sequence is followed by a c-molecule; in T'_{o_2}, the a-molecule in the looping sequence is followed by a b-molecule. Thus $T'_{o_1} \not\simeq T'_{o_2}$. However, if we compute the terms T'_{o_3} and T'_{o_4}, by applying the third and fourth occurrence respectively, with the same P_2, then we get $T'_{o_1} \simeq T'_{o_3}$ and $T'_{o_2} \simeq T'_{o_4}$. Thus, we obtain the following sets: $\mathcal{O}(R', T_o, \llbracket T'_{o_1} \rrbracket) = \{o_1, o_3\}$ and $\mathcal{O}(R', T_o, \llbracket T'_{o_2} \rrbracket) = \{o_2, o_4\}$.

The use of support-equivalence in the definition of $\mathcal{O}(R, T, \llbracket T' \rrbracket)$ allows us to consider as a single occurrence the occurrences which differ only for the support in P_2 (thus producing different, but support-equivalent, T'). As an example, in the case of T_o, the first occurrence $(a_{u_1} \mid \square \,,\ a_{u_0} \mid (b_{u_2} \cdot \widetilde{x})^L \rfloor X)$ can be produced by several concretions of the rule R' differing in their P_2 parts. In particular, admissible concretions for $\tilde{P}_2$ could be $P_2^1 = (c_{t_1} \cdot \widetilde{x})^L \rfloor X$, $P_2^2 = (c_{t_2} \cdot \widetilde{x})^L \rfloor X$,

.... However, all of them produce a single occurrence in $\mathcal{O}(R, T_o, [\![T'_{o_1}]\!])$ since, in this case, $C[P_2^i \sigma] \simeq C[P_2^j \sigma]$ for any i and j. $\qquad\square$

The following proposition holds.

Proposition 1. *Given an abstract rule* $R = (\tilde{P}_1, \tilde{P}_2, k)$ *and a concrete term* T, *let* (C, P_1) *be an occurrence of* R *in* T, *where* (P_1, P_2, k) *is a concrete rule generated by* R. *Then:*

(a) C *is determined uniquely by* P_1;
(b) P_2 *is determined uniquely by* P_1 *up to support-equivalence.*

With $T \xrightarrow{R} T'$ we denote a transition, driven by the rule $R = (\tilde{P}_1, \tilde{P}_2, k)$, from the concrete term T to the concrete term T'. We now associate a rate with transitions between concrete terms. The rate is obtained as the product of the rate k of the stochastic rewrite rule and the number of distinct occurrences of the rule within the term T resulting in T'.

Definition 12 (Rate of concrete transitions). *Given* T, T' *concrete, and an abstract reaction rule* $R = (\tilde{P}_1, \tilde{P}_2, k)$, *then* $|\mathcal{O}(R, T, [\![T']\!])|$ *is the number of distinct occurrences* (C, P_1) *of* R *in* T *resulting in* T'. *Each such occurrence is also called a* contribution *of* R *to the rate of* $T \xrightarrow{R} T'$. *The transition rate for* $T \xrightarrow{R} T'$ *is defined formally by*

$$rate_R[T, T'] \stackrel{\text{def}}{=} k \cdot |\mathcal{O}(R, T, [\![T']\!])| \ .$$

To compute the rate of an abstract transition $\tilde{T} \xrightarrow{R} \tilde{T}'$, we can just compute the rate for arbitrary concretions $T \xrightarrow{R} T'$ of that transition, because the rate is independent of the chosen concretions, i.e.:

Proposition 2. *If* $T_1 \simeq T_2$ *and* $T'_1 \simeq T'_2$, *all concrete, then, for any stochastic rewrite rule* R:
$$rate_R[T_1, T'_1] = rate_R[T_2, T'_2] \ .$$

This justifies the following definition of the abstract reaction rate.

Definition 13 (Rate of abstract transitions). *Given a stochastic rewrite rule* R, *the rate of an abstract transition* $\tilde{T} \xrightarrow{R} \tilde{T}'$ *is defined by*

$$rate_R[\tilde{T}, \tilde{T}'] \stackrel{\text{def}}{=} rate_R[T, T']$$

where T *and* T' *are arbitrary concretions of* $\tilde{T}$ *and* $\tilde{T}'$, *respectively.*

Again, an example will be helpful.

Example 3. Consider again the abstract term $\tilde{T} = a \mid a \mid b \mid b$, its concretion $T = a_{v_1} \mid a_{v_2} \mid b_{v_3} \mid b_{v_4}$ and the rule $R = a \mid b \stackrel{k}{\mapsto} c$. In Example 2 we have defined the set of occurrences $\mathcal{O}(R, T, [\![T']\!])$ which contains exactly four elements, thus

$rate_R[T, T'] = k \cdot 4$. As a consequence, for the abstract term $\tilde{T}' = a \,|\, c \,|\, b$ we obtain $rate_R[\tilde{T}, \tilde{T}'] = k \cdot 4$.

Similarly, for the abstract term $\tilde{T}_o = a \,|\, a \,|\, (b{\cdot}c{\cdot}b{\cdot}a)^L \rfloor \epsilon$, given its concretion T_o and the stochastic rewrite rule R' defined in Example 2, we get the following:

- $\tilde{T}'_{o_1} = a \,|\, (c{\cdot}c{\cdot}b{\cdot}a)^L \rfloor \epsilon$ is the abstraction of T'_{o_1};
- $\tilde{T}'_{o_2} = a \,|\, (c{\cdot}a{\cdot}b{\cdot}c)^L \rfloor \epsilon$ is the abstraction of T'_{o_2}.

We can then derive the following rates: $rate_{R'}[\tilde{T}_o, \tilde{T}'_{o_1}] = rate_{R'}[\tilde{T}_o, \tilde{T}'_{o_2}] = k' \cdot 2$ since $|\mathcal{O}(R', T_o, [\![T'_{o_1}]\!])| = |\mathcal{O}(R', T_o, [\![T'_{o_2}]\!])| = 2$. $\qquad\square$

We can now unroll the definitions of occurrences and transition rates to get the semantics of SCLS. The stochastic transition system for abstract terms is defined as follows.

Definition 14 (Semantics). *Given a finite set $\mathcal{R}$ of stochastic rewrite rules, the semantics of SCLS is the least labelled transition relation satisfying the following rule:*

$$\frac{R = \tilde{P}_1 \overset{k}{\mapsto} \tilde{P}_2 \in \mathcal{R} \quad (C, P_1) \in \mathcal{O}(R, T, [\![T']\!]) \quad T \equiv C[P_1\sigma] \quad T' \doteq C[P_2\sigma]}{\tilde{T} \xrightarrow{R, k \cdot |\mathcal{O}(R, T, [\![T']\!])|} \tilde{T}'}$$

The stochastic reduction semantics associates with each transition a rate which is the parameter of an exponential distribution that characterizes the stochastic behaviour of the activity corresponding to the applied rewrite rule. As we have already noticed, the rate is obtained as the product of the rewrite rate constant and the number of occurrences of the rule within the starting term (thus counting the exact number of reactants to which the rule can be applied and which produce the same result).

Our stochastic semantics is essentially a *Continuous Time Markov Chain* (CTMC). We can follow a standard simulation procedure that corresponds to Gillespie's simulation algorithm [14]. We have developed a prototype simulator (called SCLSm) for SCLS in the language F#.

In the next two sections we report some experimental results. We used concrete terms and patterns just for defining a consistent methodology for counting the occurrences of a rule within a term. When modelling, however, we would like to reason in an abstract way and we are not interested in concrete patterns or terms anymore. Thus, to lighten the notation in the next two sections, we resort again to the plain convention (without the tilde) to denote abstract patterns and terms.

4   Modelling the Lactose Operon

To show that SCLS can be easily used to model and simulate cellular pathways, we give a SCLS model of the well-known regulation process of the lactose operon in *Escherichia coli* and we use our prototype simulator to analyze the process in different situations.

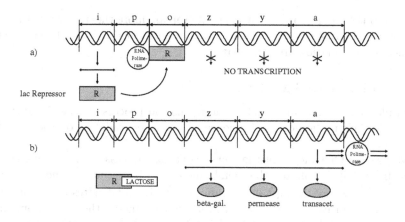

Fig. 2. The regulation process in the Lac Operon

E. coli is a bacterium often present in the intestine of many animals. It is one of the most completely studied of all living things and it is a favorite organism for genetic engineering. Cultures of E. coli can be made to produce unlimited quantities of the product of an introduced gene. As most bacteria, E.coli is often exposed to a constantly changing physical and chemical environment, and reacts to changes in its environment through changes in the kinds of enzymes it produces. In order to save energy, bacteria do not synthesize degradative enzymes unless the substrates for these enzymes are present in the environment. For example, E. coli does not synthesize the enzymes that degrade lactose unless lactose is in the environment. This result is obtained by controlling the transcription of some genes into the corresponding enzymes.

Two enzymes are involved in lactose degradation: the *lactose permease*, which is incorporated in the membrane of the bacterium and actively transports the sugar into the cell, and the *beta galactosidase*, which splits lactose into glucose and galactose. The bacterium produces also the *transacetylase* enzyme, whose role in the lactose degradation is marginal.

The sequence of genes in the DNA of E. coli which produces the described enzymes, is known as the *lactose operon*.

The first three genes of the operon (i, p and o) regulate the production of the enzymes, and the last three (z, y and a), called *structural genes*, are transcribed (when allowed) into the mRNA for beta galactosidase, lactose permease and transacetylase, respectively.

The regulation process is as follows (see Figure 2): gene i encodes the *lac Repressor*, which, in the absence of lactose, binds to gene o (the *operator*). Transcription of structural genes into mRNA is performed by the RNA polymerase enzyme, which usually binds to gene p (the *promoter*) and scans the operon from left to right by transcribing the three structural genes z, y and a into a single mRNA fragment. When the lac Repressor is bound to gene o, it becomes an obstacle for the RNA polymerase, and the transcription of the structural genes is not performed. On the other hand, when lactose is present inside the bacterium, it binds to the Repressor and this cannot stop anymore the activity of the RNA polymerase. In this case the transcription is performed and the three enzymes for lactose degradation are synthesized.

4.1 Stochastic CLS Model

A detailed mathematical model of the regulation process can be found in [34]. It includes information on the influence of lactose degradation on the growth of the bacterium.

We give a SCLS model of the gene regulation process, with stochastic rates taken from [33]. We model the membrane of the bacterium as the looping sequence $(m)^L$, where the alphabet symbol m generically denotes the whole membrane surface in normal conditions. Moreover, we model the lactose operon as the sequence $lacI \cdot lacP \cdot lacO \cdot lacZ \cdot lacY \cdot lacA$ ($lacI$–A for short), in which each symbol corresponds to a gene. We replace $lacO$ with RO in the sequence when the lac Repressor is bound to gene o, and $lacP$ with PP when the RNA polymerase is bound to gene p. When the lac Repressor and the RNA polymerase are unbound, they are modelled by the symbols $repr$ and $polym$, respectively. We model the mRNA of the lac Repressor as the symbol $Irna$, a molecule of lactose as the symbol $LACT$, and beta galactosidase, lactose permease and transacetylase enzymes as symbols $betagal, perm$ and $transac$, respectively. Finally, since the three structural genes are transcribed into a single mRNA fragment, we model such mRNA as a single symbol Rna.

The initial state of the bacterium when no lactose is present in the environment and when 100 molecules of lactose are present are modelled, respectively, by the following terms (where $n \times T$ stands for a parallel composition $T \,|\, \ldots \,|\, T$ of length n):

$$Ecoli \; ::= \; (m)^L \,\rfloor\, (lacI\text{–}A \,|\, 30 \times polym \,|\, 100 \times repr) \qquad (1)$$

$$EcoliLact \; ::= \; Ecoli \,|\, 100 \times LACT \qquad (2)$$

The transcription of the DNA, the binding of the lac Repressor to gene o, and the interaction between lactose and the lac Repressor are modelled by the following set of stochastic rewrite rules:

$$lacI \cdot \widetilde{x} \xmapsto{0.02} lacI \cdot \widetilde{x} \,|\, Irna \qquad Irna \xmapsto{0.1} Irna \,|\, repr \qquad \text{(R1-R2)}$$

$$polym \,|\, \widetilde{x} \cdot lacP \cdot \widetilde{y} \xmapsto{0.1} \widetilde{x} \cdot PP \cdot \widetilde{y} \qquad \text{(R3)}$$

$$\widetilde{x} \cdot PP \cdot \widetilde{y} \xmapsto{0.01} polym \,|\, \widetilde{x} \cdot lacP \cdot \widetilde{y} \qquad \text{(R4)}$$

$$\widetilde{x} \cdot PP \cdot lacO \cdot \widetilde{y} \xmapsto{20.0} polym \,|\, Rna \,|\, \widetilde{x} \cdot lacP \cdot lacO \cdot \widetilde{y} \qquad \text{(R5)}$$

$$Rna \xmapsto{0.1} Rna \,|\, betagal \,|\, perm \,|\, transac \qquad \text{(R6)}$$

$$repr \,|\, \widetilde{x} \cdot lacO \cdot \widetilde{y} \xmapsto{1.0} \widetilde{x} \cdot RO \cdot \widetilde{y} \qquad \widetilde{x} \cdot RO \cdot \widetilde{y} \xmapsto{0.01} repr \,|\, \widetilde{x} \cdot lacO \cdot \widetilde{y} \qquad \text{(R7-R8)}$$

$$repr \,|\, LACT \xmapsto{0.005} RLACT \qquad RLACT \xmapsto{0.1} repr \,|\, LACT \qquad \text{(R9-R10)}$$

Rules (R1) and (R2) describe the transcription and translation of gene i into the lac Repressor (assumed for simplicity to be performed without the intervention of the RNA polymerase). Rules (R3) and (R4) describe binding and unbinding of the RNA polymerase to gene p. Rules (R5) and (R6) describe the transcription and translation of the three structural genes. Transcription of such genes can be performed only when the sequence contains $lacO$ instead of RO, that is when the lac Repressor is not bound to gene o. Rules (R7) and (R8) describe binding and unbinding of the lac Repressor to gene o. Finally, rules (R9) and (R10) describe the binding and unbinding, respectively, of the lactose to the lac Repressor. The following rules describe the behaviour of the three enzymes for lactose degradation:

$$(\widetilde{x})^L \rfloor (perm \,|\, X) \overset{0.1}{\mapsto} (perm \cdot \widetilde{x})^L \rfloor X \qquad (\text{R11})$$

$$LACT \,|\, (perm \cdot \widetilde{x})^L \rfloor X \overset{0.001}{\mapsto} (perm \cdot \widetilde{x})^L \rfloor (LACT | X) \qquad (\text{R12})$$

$$betagal \,|\, LACT \overset{0.001}{\mapsto} betagal \,|\, GLU \,|\, GAL \qquad (\text{R13})$$

Rule (R11) describes the incorporation of the lactose permease in the membrane of the bacterium, rule (R12) the transportation of lactose from the environment to the interior performed by the lactose permease, and rule (R13) the decomposition of the lactose into glucose (denoted GLU) and galactose (denoted GAL) performed by the beta galactosidase.

The following rules describe the degradation of all the proteins and pieces of mRNA involved in the process:

$$perm \overset{0.001}{\mapsto} \epsilon \qquad Irna \overset{0.001}{\mapsto} \epsilon \qquad transac \overset{0.001}{\mapsto} \epsilon \qquad (\text{R14-R16})$$

$$repr \overset{0.002}{\mapsto} \epsilon \qquad betagal \overset{0.01}{\mapsto} \epsilon \qquad Rna \overset{0.01}{\mapsto} \epsilon \qquad (\text{R17-R19})$$

$$RLACT \overset{0.002}{\mapsto} LACT \qquad (perm \cdot \widetilde{x})^L \rfloor X \overset{0.001}{\mapsto} (\widetilde{x})^L \rfloor X \qquad (\text{R20-R21})$$

We recall that sequences are not allowed as context of application of rules, hence rule (R14) cannot be applied to $perm$ when this is an element of the looping sequence representing the membrane of the bacterium. This motivates the presence of the rule (R21).

4.2 Simulation Results

We simulated the evolution of the bacterium in the absence of lactose (modelled by the term $Ecoli$ of Equation (1)) and in the presence of 100 molecules of lactose in the environment (modelled by the term $EcoliLact$ of Equation (2)).

In Figure 3 we show the results of the two simulations. The first graph shows that in the absence of lactose the production of the beta galactosidase and lactose permease enzymes starts after more than 750 seconds, and that the number of such enzymes is always smaller than 20. Moreover, this graph shows that the lactose permeases, once produced, become immediately part of the membrane of the bacterium, because the number of such enzymes not on the membrane remains always small.

The second and the third graphs show the results of the simulation when the lactose is present in the environment. In this case the production of the enzymes starts almost immediately (as shown by the second graph), but we

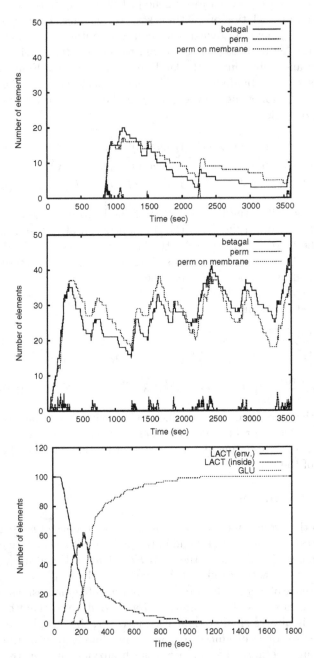

Fig. 3. Simulation results: production of enzymes in the absence (top) and presence (middle) of lactose, and degradation of lactose into glucose (bottom)

remark that the fact that the starting times in the production of enzymes in the two simulations are different is not relevant. The amount of time elapsed before the production of these enzymes does not depend on the presence of the lactose in the environment, as the lactose cannot enter the bacterium until some molecules of permease have joint the membrane.

Finally, the third graph shows that if lactose is present in the environment, it starts entering the bacterium after some molecules of lactose permease join the membrane and it is decomposed into glucose and galactose. Once some molecules of lactose permease join the membrane, the lactose starts entering the bacterium (see the third graph). In fact, the third graph shows that the number of molecules in the environment rapidly decreases.

Once entered the bacterium, the lactose interacts with the lac Repressor, and this favors the production of more enzymes.

Once all the molecules of lactose have been decomposed, the number of lac Repressors increases, reaching the same values of the first simulation. The number of beta galactosidase and lactose permease enzymes, instead, does not decrease, and hence does not reach the values of the first simulation. This happens because the degradation of such enzymes, and of the mRNA from which they are translated, is a very slow process, which would take much more time than the time of the simulations we performed.

Notice that the curves proposed in the figures (and the ones in the next section) result from a single experiment chosen among several we have performed with the same initial conditions. Even though the outputs of probabilistic simulations are inherently approximations of the overall behaviour of a system, in our cases the experiments gave always the same results, apart for negligible fluctuations. In tens of simulations no rare events manifested. Thus, instead of computing an average of the different simulations we decided to choose just one of them as a good representative of the standard behaviour of the system. Thus, we are confident that the proposed simulations reflect a realistic behaviour of the overall system.

5 Modelling Quorum Sensing

Traditionally, bacteria have been studied as independent individuals. Now, it is recognised that many bacteria have the ability of monitoring their population density and modulating their gene expressions according to this density. This process is called *quorum sensing*.

The process of quorum sensing consists in two activities, one involving one or more diffusible small molecules (called *autoinducers*) and the other involving one or more transcriptional activator proteins (*R-proteins*) located within the cell. The autoinducer can cross the cellular membrane, and thus it can diffuse either out or in bacteria.

The production of the autoinducer is regulated by the R-protein. The R-protein by itself is not active without the corresponding autoinducer. The autoinducer molecule can bind to the R-protein to form an *autoinducer/R-protein* complex, which binds to a target of the DNA sequence enhancing the transcription of specific genes. Usually, these genes regulate both the production of specific behavioural traits (as we will show in the following) and the production of the autoinducer and of the R-protein.

At low cell density, the autoinducer is synthesized at basal levels and diffuse in the environment where it is diluted. With high cell density both the extracellular and intracellular concentrations of the autoinducer increase until they reach thresholds beyond which the autoinducer is produced autocatalytically. The autocatalytic production results in a dramatic increase of product concentration.

Quorum sensing behaviour is very widespread in bacteria. An example is the regulation of the bioluminescence in the symbiotic marine bacterium *Vibrio fischeri*, which colonizes the light organs of marine fishes and squids. The bacteria only luminesce when they are found in high concentrations in the light organs, while they do not emit light when they are free swimming [30]. Another example is given by the bacterium *Pseudomonas aeruginosa*, a prevalent human pathogen [31]. The ability of *P. aeruginosa* to infect a host mainly is based on controlling its virulence by quorum sensing. The level of virulence expressed by isolated bacteria is very low, thus avoiding host response. When a colony has reached a certain density, the production of virulence factors is autoinduced by quorum sensing, and it is generally sufficient to overcome the defenses of the host.

The quorum sensing system of *P. aeruginosa* has two regulatory systems regulating the expression of elastase LasA and elastase LasB, respectively. The two enzymes are responsible for pulmonary hemorrhages associated with *P. aeruginosa* infections. In this paper we are interested in the regulatory system of elastase LasB, named the *las* system.

A schematic description of the *las system* is as follows (arrows from an element to another one represent the production of the second element starting form the first one, arrows with the ++ label represent catalysed faster productions):

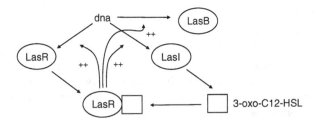

The autoinducer 3-oxo-C12-HSL and the transcriptional activator protein LasR are produced at basal rates starting from the dna and LasI, respectively. The LasR/3-oxo-C12-HSL dimer, which is the activated form of LasR, promotes the production of itself, of the autoinducer and of the LasB enzyme. The formation of the dimer is controlled mainly by the concentration of the autoinducer, which is influenced by the number of bacteria.

5.1   Stochastic CLS Model

We now give the SCLS model of the quorum sensing process. We do not model the production of the LasB as it has not an active role in the regulation process. The initial state of each bacterium is:

$$Bact \ ::= \ (m)^L \,\rfloor\, (lasO \cdot lasR \cdot lasI)$$

where the looping sequence $(m)^L$ represents the bacterium membrane, $lasO$ the target of the DNA sequence where LasR/3-oxo-C12-HSL complex binds to for promoting DNA transcription, and $lasR$ and $lasI$ the genes that encode $LasR$ and the autoinducer.

This model shows one of the advantages of using terms for describing the structure of biological systems in SCLS. In fact, in order to model a population of n bacteria we have to describe only one bacterium, and then compose n copies of such a description by using the parallel composition operator. In other words, we model a population of n bacteria simply as $n \times Bact$.

We now give the stochastic rewrite rules describing the protein/protein and protein/DNA interactions in the described systems. Again, we have only to give the rules for one bacterium, and they will be applicable in all the n bacteria of the considered population.

$$
\begin{array}{rcll}
lasO \cdot lasR \cdot lasI & \overset{20}{\longmapsto} & lasO \cdot lasR \cdot lasI \,|\, LasR & \text{(R1)} \\[4pt]
lasO \cdot lasR \cdot lasI & \overset{5}{\longmapsto} & lasO \cdot lasR \cdot lasI \,|\, LasI & \text{(R2)} \\[4pt]
LasI & \overset{8}{\longmapsto} & LasI \,|\, 3oxo & \text{(R3)} \\[4pt]
3oxo \,|\, LasR \ \overset{0.25}{\longmapsto}\ 3R & \quad 3R \overset{400}{\longmapsto} 3oxo \,|\, LasR & & \text{(R4-R5)} \\[4pt]
3R \,|\, lasO \cdot lasR \cdot lasI & \overset{0.25}{\longmapsto} & 3RO \cdot lasR \cdot lasI & \text{(R6)} \\[4pt]
3RO \cdot lasR \cdot lasI & \overset{10}{\longmapsto} & 3R \,|\, lasO \cdot lasR \cdot lasI & \text{(R7)} \\[4pt]
3RO \cdot lasR \cdot lasI & \overset{1200}{\longmapsto} & 3RO \cdot lasR \cdot lasI \,|\, LasR & \text{(R8)} \\[4pt]
3RO \cdot lasR \cdot lasI & \overset{300}{\longmapsto} & 3RO \cdot lasR \cdot lasI \,|\, LasI & \text{(R9)} \\[4pt]
(m)^L \rfloor (3oxo \,|\, X) & \overset{30}{\longmapsto} & 3oxo \,|\, (m)^L \rfloor X & \text{(R10)} \\[4pt]
3oxo \,|\, (m)^L \rfloor X & \overset{1}{\longmapsto} & (m)^L \rfloor (3oxo \,|\, X) & \text{(R11)} \\[4pt]
LasI \overset{1}{\longmapsto} \epsilon \quad\quad LasR & \overset{1}{\longmapsto} \epsilon \quad\quad 3oxo \overset{1}{\longmapsto} \epsilon & & \text{(R12-S14)}
\end{array}
$$

Rules (R1) and (R2) describe the production from the DNA of proteins LasR and LasI, respectively. For the sake of simplicity we do not model the transcription of the DNA into mRNA. Rule (R3) describes the production of the autoinducer 3-oxo-C12-HSL, denoted $3oxo$, performed by the LasI enzyme. Rules (R4) and (R5) describe the complexation and decomplexation of the autoinducer and the LasR protein, where the complex is denoted $3R$. Rules from (R7) to (R9) describe the binding of the activated autoinducer to the DNA and its influence in the production of LasR and LasI. Rules (R10) and (R11) describe the autoinducer exiting and entering the bacterium. The kinetic constants associated with these two rules give a measure of the autoinducer dilution. Finally, rules from (R12) to (R14) describe the degradation of proteins.

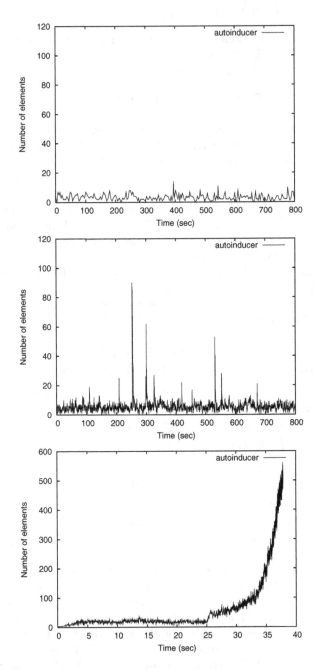

Fig. 4. Simulation results: quantity of autoinducer inside one bacterium in a population of one (top), five (middle) and twenty (bottom) bacteria

5.2   Experimental Results

We simulated the behaviour of a population of *P. aeruginosa* by varying the number of individuals. In Figure 4 we show how the concentration of the autoinducer varies inside bacteria when the population is composed by one, five and twenty individuals. In the last two cases we show the autoinducer concentration inside one only bacterium (the concentrations inside the others are analogous).

When the number of bacteria increases also the concentration of the autoinducer in the extracellular space increases. As a consequence the concentration of the autoinducer in the intracellular spaces increases as well and the quorum sensing process starts. Note that the kinetic constants of rules (R10) and (R11), regulating the autoinducer exiting and entering the membrane, cause the bacteria to maintain the autoinducer production mostly at a basal rate when the population size is one or five. When the population size is twenty the quorum sensing starts after a few seconds thus causing a very high autocatalytic autoinducer production. Increasing the ratio between the kinetic constants of (R10) and (R11) would cause the quorum sensing to be triggered when the number of individuals is bigger.

6   Conclusions

As we have seen, SCLS allows representing membranes and operations on them. Other formalisms were developed to describe membrane systems. Among them we cite Brane Calculi [7] and P-Systems [21].

SCLS can describe situations that cannot be easily captured by the above mentioned formalisms, which consider membranes as atomic objects. An example of this is given by the representation of the membrane of *Escherichia coli*, shown in Section 4. Representing the membrane as a sequence of elements permits the definition of different functionalities depending on the type and the number of elements on the membrane itself. In the example, the presence and the number of lactose permeases on the bacterium membrane regulates the transportation of lactose inside the bacterium. Moreover, as no restrictions are imposed on the format of rewrite rules, SCLS seems to be suitable for the description of a wider class of systems than the one the formalisms mentioned above easily handle.

Quorum sensing is a complex biological process, which is not based on signals and receptors but only on the concentration of a protein freely crossing bacteria membranes. Many mathematical models have been developed for describing this challenging phenomenon [13,16,32]. These models consider various aspects of the problem: the diffusion of the autoinducer, its degradation, the percentage of "up-regulated" bacteria (the ones with an enhanced production of the autoinducer), the density of bacteria, their size, etc. However, all these models describe the process at a very abstract level. They consider that the intracellular concentration of the autoinducer is a function of the density of the bacteria, although modulated by other factors. Thus they start from this assumption to study the behaviour of the system with different values of the parameters.

The SCLS model is based on a different approach. A single bacterium is described by means of a set of rewrite rules modelling its internal processes. These rules also model the autoinducer crossing (in both directions) the cellular membranes and the autoinducer degrading at the same rate both inside and outside cells.

Differently from the mathematical models mentioned above, the SCLS model describes the elementary processes each bacterium performs, and the quorum sensing results from the activity of a sufficient number of bacteria. The stochastic nature of SCLS allows observing fluctuations of the autoinducer concentration which in the first two cases considered are not sufficient to trigger quorum sensing. Moreover, our model shows the discrete behaviour of the binding between the autoinducer/R-protein complex and DNA. In Figures 5 and 6 we show the dynamics of the binding between the autoinducer/R-protein complex and the DNA in one of the bacteria of the populations of five and twenty bacteria, respectively. In Figure 5, since the autoinducer and the R protein are produced at basal levels, the concentration of the autoinducer/R-protein complex inside the bacterium is low. As a consequence, the complex does not bind to DNA, except in the few cases of stochastic peaks in the autoinducer concentration. In Figure 6, when the quorum sensing process starts (in this case approximately after 30 seconds), the concentration of the autoinducer/R-protein complex increases sharply causing the complex to be constantly bound to the DNA. As a consequence, the autoinducer is produced autocatalitically.

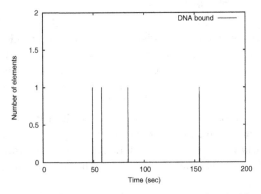

Fig. 5. The binding of the autoinducer/R-protein complex inside a bacterium in the population of five

6.1   Related Work

Cell biology, the study of the morphological and functional organization of cells, is now an established field in biochemical research. Computer Science can help the research in cell biology in several ways. For instance, it can provide biologists with models and formalisms capable of describing and analyzing complex systems such as cells.

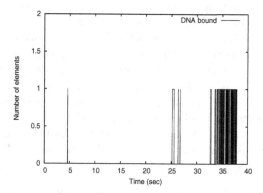

Fig. 6. The binding of the autoinducer/R-protein complex inside a bacterium in the population of twenty

Qualitative Models. In the last few years many formalisms originally developed by computer scientists to model systems of interacting components have been applied to Biology. Among these, there are Petri Nets [19], Hybrid Systems [1], and the π-calculus [10,29]. Moreover, new formalisms have been defined for describing biomolecular and membrane interactions [4,7,8,12,24,26]. Others, such as P-Systems [21], have been proposed as biologically inspired computational models and have been later applied to the description of biological systems.

The π-calculus and new calculi based on it [24,26] have been particularly successful in the description of biological systems, as they allow describing systems in a compositional manner. Interactions of biological components are modelled as communications on channels whose names can be passed; sharing names of private channels allows describing biological compartments.

These calculi offer very low-level interaction primitives, but may cause the description models to become very large and difficult to read. Calculi such as those proposed in [7,8,12] give a more abstract description of systems and offer special biologically motivated operators. However, they are often specialized to the description of some particular kinds of phenomena such as membrane interactions or protein interactions.

P-Systems [21] have a simple notation and are not specialized to the description of a particular class of systems, but they are still not completely general. For instance, it is possible to describe biological membranes and the movement of molecules across membranes, and there are some variants able to describe also more complex membrane activities. However, the formalism is not so flexible to allow describing easily new activities observed on membranes without extending the formalism to model such activities.

Danos and Laneve [12] proposed the κ-calculus. This formalism is based on graph rewriting where the behaviour of processes (compounds) and of set of processes (solutions) is given by a set of rewrite rules which account for, e.g., activation, synthesis and complexation by explicitly modelling the binding sites of a protein.

The Calculus of Looping Sequences presented in the present paper, has no explicit way to model protein domains (however they can be encoded, and a variant with explicit binding has been defined in [3]), but accounts for an explicit mechanism (the *looping sequences*) to deal with compartments and membranes. Thus, while the κ-calculus seems more suitable to model protein interactions, CLS allows for a more natural description of membrane interactions.

Another feature lacking in other formalisms is the capacity to express ordered sequences of elements. To the best of our knowledge, CLS is the first formalism offering such a feature in an explicit way, thus allowing to naturally operate over proteins or DNA fragments which should be frequently defined as ordered sequences of elements.

Stochastic Models. Among stochastic process algebras we would like to mention the stochastic extension of the π-calculus, given by Priami et al. in [25], and the PEPA framework proposed by Hillston in [15].

The stochastic engine behind PEPA and the Stochastic π-calculus is constructed on the intuition of cooperating agents under different bandwidth limits. If two agents are interacting, the time spent for a communication is given by the slowest of the agents involved. Differently, our stochastic semantics is defined in terms of the collision-based paradigm introduced by Gillespie. A similar approach is taken in the quantitative variant of the κ-calculus ([11]) and in BioSPi ([25]). Motivated by the law of mass action, here we need to count the number of the reactants present in a system in order to compute the exact rate of a reaction. We already mentioned the work by Krivine et al. [17], in which a stochastic semantics for bigraphs has been developed. The intuitive and natural methodology they have developed to count rule occurrences has been adapted, in the present paper, to count the number of occurrences of stochastic rewrite rules within CLS terms.

An alternative stochastic semantics for CLS has been defined in [2]. Such a semantics computes the transition rates in a compositional way and it is rather complicated. Moreover, in [20] and in preliminary versions of the present paper we defined the stochastic extension of CLS (upon which the current version of our simulator is based) by enriching rewrite rules with rate functions rather than rate constants. Such functions allow the definition of kinetics that are more complex than the standard mass-action ones, but are rather difficult to be used when modelling systems. By following the approach of [17], in this paper we have been able to give a natural and simpler definition of SCLS.

We would also like to mention that a computational model describing the quorum sensing process in *Vibrio fischeri* [30] by modelling single bacteria is presented in [22]. The model is defined with a variant of P-Systems.

Tools and Applications. Among the simulation tools based on other stochastic formalisms we mention the following ones:

- SPiM (http://research.microsoft.com/~aphillip/spim/) and Cytosim (http://www.cosbi.eu/Rpty_Soft_CytoSim.php) are constructed on the Stochastic π-calculus;

- the Beta Workbench (`http://www.cosbi.eu/Rpty_Soft_BetaWB.php`) provides a collection of tools based on Beta-binders [24];
- Bio-PEPA (`http://www.dcs.ed.ac.uk/home/stg/software/biopepa`) is a biologically inspired extension of PEPA;
- PSym (`http://psystems.disco.unimib.it/`) is a simulation tool developed for P-Systems.

These tools, as our prototype simulator for SCLS, can be used to perform stochastic simulations. Other tools often used to study biological systems, such as GEPASI (`http://www.gepasi.org/`), simulate systems by solving differential equations. This kind of simulation is well-suited for chemical solutions with big quantities of reactants, and becomes less precise when the number of reactants decreases (as in cells) as it is based on a continuous representation of the quantities of reactants.

In conclusion, we also point out that the translation of SBML descriptions of cellular pathways (`http://sbml.org`) into SCLS is trivial, provided the reactions in the pathways are based on standard mass action kinetics.

References

1. Alur, R., Belta, C., Ivancic, F., Kumar, V., Mintz, M., Pappas, G.J., Rubin, H., Schug, J.: Hybrid modeling and simulation of biomolecular networks. In: Di Benedetto, M.D., Sangiovanni-Vincentelli, A.L. (eds.) HSCC 2001. LNCS, vol. 2034, pp. 19–32. Springer, Heidelberg (2001)
2. Barbuti, R., Caravagna, G., Maggiolo-Schettini, A., Milazzo, P., Pardini, G.: The Calculus of Looping Sequences. In: Bernardo, M., Degano, P., Zavattaro, G. (eds.) SFM 2008. LNCS, vol. 5016, pp. 387–423. Springer, Heidelberg (2008)
3. Barbuti, R., Maggiolo-Schettini, A., Milazzo, P.: Extending the Calculus of Looping Sequences to Model Protein Interaction at the Domain Level. In: Măndoiu, I.I., Zelikovsky, A. (eds.) ISBRA 2007. LNCS (LNBI), vol. 4463, pp. 638–649. Springer, Heidelberg (2007)
4. Barbuti, R., Maggiolo-Schettini, A., Milazzo, P., Troina, A.: A calculus of looping sequences for modelling microbiological systems. Fund. Inform. 72, 21–35 (2006)
5. Barbuti, R., Maggiolo-Schettini, A., Milazzo, P., Troina, A.: Bisimulation congruences in the calculus of looping sequences. In: Barkaoui, K., Cavalcanti, A., Cerone, A. (eds.) ICTAC 2006. LNCS, vol. 4281, pp. 93–107. Springer, Heidelberg (2006)
6. Barbuti, R., Maggiolo-Schettini, A., Milazzo, P., Troina, A.: Bisimulations in Calculi Modelling Membranes. Formal Aspects of Computing (to appear, 2008)
7. Cardelli, L.: Brane calculi. Interactions of biological membranes. In: Danos, V., Schachter, V. (eds.) CMSB 2004. LNCS (LNBI), vol. 3082, pp. 257–280. Springer, Heidelberg (2005)
8. Chabrier-Rivier, N., Chiaverini, M., Danos, V., Fages, F., Schachter, V.: Modeling and querying biomolecular interaction networks. Theor. Comput. Sci. 325, 25–44 (2004)
9. Ciocchetta, F., Hillston, J.: Bio-PEPA: a framework for the modelling and analysis of biological systems. Theor. Comput. Sci. (to appear, 2008)
10. Curti, M., Degano, P., Priami, C., Baldari, C.T.: Modelling Biochemical Pathways through Enhanced pi-calculus. Theor. Comput. Sci. 325, 111–140 (2004)

11. Danos, V., Feret, J., Fontana, W., Krivine, J.: Scalable modelling of biological pathways. In: Shao, Z. (ed.) APLAS 2007. LNCS, vol. 4807, pp. 139–157. Springer, Heidelberg (2007)
12. Danos, V., Laneve, C.: Formal molecular biology. Theor. Comput. Sci. 325, 69–110 (2004)
13. Dockery, J.D., Keener, J.P.: A mathematical model for quorum sensing in Pseudomonas aeruginosa. Bulletin of Mathematical Biology 63, 95–116 (2001)
14. Gillespie, D.: Exact stochastic simulation of coupled chemical reactions. J. Phys. Chem. 81, 2340–2361 (1977)
15. Hillston, J.: A Compositional Approach to Performance Modelling. Cambridge University Press, Cambridge (1996)
16. James, S., Nilsson, P., James, G., Kjelleberg, S., Fagerstroem, T.: Luminescence Control in the Marine Bacterium Vibrio fischeri: An Analysis of the Dynamics of lux Regulation. Journal of Molecular Biology 296, 1127–1137 (2000)
17. Krivine, J., Milner, R., Troina, A.: Stochastic Bigraphs. In: Proc. of the 24th Conference on Mathematical Foundations of Programming Semantics (MFPS 2008). ENTCS. Elsevier, Amsterdam (to appear, 2008)
18. Kwiatkowska, M., Norman, G., Parker, D.: Probabilistic symbolic model checking with PRISM: a hybrid approach. Int. J. on Software Tools for Technology Transfer 6, 128–142 (2004)
19. Matsuno, H., Doi, A., Nagasaki, M., Miyano, S.: Hybrid Petri net representation of gene regulatory network. In: Prooceedings of Pacific Symposium on Biocomputing, pp. 341–352. World Scientific Press, Singapore (2000)
20. Milazzo, P.: Qualitative and quantitative formal modeling of biological systems. Ph.D. Thesis, University of Pisa (2007)
21. Păun, G.: Membrane computing. An introduction. Springer, Heidelberg (2002)
22. Pérez-Jiménez, M.J., Romero-Campero, F.J.: Modelling Vibrio fischeri's behaviour using P systems. In: Proc. of the 8th European Conference on Artificial Life, Systems Biology Workshop (2005)
23. Priami, C.: Stochastic π-calculus. The Computer Journal 38, 578–589 (1995)
24. Priami, C., Quaglia, P.: Beta Binders for Biological Interactions. In: Danos, V., Schachter, V. (eds.) CMSB 2004. LNCS (LNBI), vol. 3082, pp. 20–33. Springer, Heidelberg (2005)
25. Priami, C., Regev, A., Shapiro, E., Silverman, W.: Application of a stochastic name-passing calculus to representation and simulation of molecular processes. Inform. Process. Lett. 80, 25–31 (2001)
26. Regev, A., Panina, E.M., Silverman, W., Cardelli, L., Shapiro, E.: BioAmbients: an abstraction for biological compartments. Theor. Comput. Sci. 325, 141–167 (2004)
27. Regev, A., Shapiro, E.: Cells as computation. Nature 419, 343 (2002)
28. Regev, A., Shapiro, E.: The π-calculus as an abstraction for biomolecular systems. In: Modelling in Molecular Biology. Natural Computing Series, pp. 219–266. Springer, Heidelberg (2004)
29. Regev, A., Silverman, W., Shapiro, E.Y.: Representation and simulation of biochemical processes using the pi-calculus process algebra. In: Proc. of Pacific Symposium on Biocomputing, pp. 459–470. World Scientific Press, Singapore (2001)
30. Stevens, A.M., Greenberg, E.P.: Quorum sensing in Vibrio fischeri: essential elements for activation of the luminescence genes. J. of Bacteriology 179, 557–562 (1997)
31. Van Delden, C., Iglewski, B.H.: Cell-to-cell signaling and Pseudomonas aeruginosa infections. Emerg. Infect. Dis. 4, 551–560 (1998)

32. Ward, J.P., King, J.R., Koerber, A.J., Croft, J.M., Sockett, R.E., Williams, P.: Early development and quorum sensing in bacterial biofilms. Journal of Mathematical Biology 47, 23–55 (2003)
33. Wilkinson, D.: Stochastic modelling for Systems Biology. Chapman & Hall/CRC, Boca Raton (2006)
34. Wong, P., Gladney, S., Keasling, J.D.: Mathematical model of the lac operon: inducer exclusion, catabolite repression, and diauxic growth on glucose and lactose. Biotechnology Progress 13, 132–143 (1997)

The BlenX Language with Biological Transactions

Federica Ciocchetta

Laboratory for Foundations of Computer Science, The University of Edinburgh,
Edinburgh EH9 3JZ, Scotland

Abstract. An extension of the BlenX language with *biological transactions*, called TBlenX, is presented. The aim of this extension is to model a sequence of elementary actions as if it were atomic. This extension is useful when we need to specify multi-reactant multi-product reactions or when we use a sequence of actions to represent a biological interaction. Some properties of these transactions are discussed and some examples are reported to illustrate our extension.

Keywords: Systems biology, process algebras, biological transactions, multiple-reactant multiple-product reactions.

1 Introduction

In the last years process algebras have been used to model and analyse biological systems [38,34,15,9,4,12]. These techniques have been originally defined in computer science for the analysis of complex concurrent systems and they seem to be appropriate for representing biological systems as well. One example is the π-calculus with its stochastic version [30,29,26]. Moreover, there have been some efforts to define specific calculi for biology [37,33,31].

Recently, BlenX, a language based on Beta-binders [33], has been defined and implemented [17,19,18]. The language is based on the concept of *boxes*, equipped with some sites (*binders*) and with π-like processes (*processes*) inside. The boxes abstract biological entities, whereas the binders express their interaction capabilities and processes handle the manipulation of the binders and drive the internal behaviour of the boxes in which processes are. BlenX allows us to represent some biological phenomena, such as the join between two bio-processes, the split of one bio-process into two, the change of the bio-process interface by hiding, unhiding and exposing a site. The interaction between two boxes can happen if they have compatible interaction sites and not identical channel names as it happens in classical process calculi. The definition of *compatibility* is expressed by an affinity function applied to the types of the sites involved in the communication. This new form of communication is relevant in biology, where interactions happen on the basis of sensitivity between active sites of entities rather than on the basis of exact complementarity.

Further notions of BlenX with respect to Beta-binders are the creation, the deletion, the complexation and the decomplexation of boxes. The notion of *complex*, resulted from the complexation action, is introduced to describe a set of boxes physically bound together. BlenX offers a formal and efficient definition of joins and splits by means of the *event* construct. A single *event* is defined by a condition with some associated

C. Priami (Ed.): Trans. on Comput. Syst. Biol. IX, LNBI 5121, pp. 114–152, 2008.

actions. Conditions are global and whenever they become true the associated actions are enabled. The two levels of control in the language (interactions and events) represent the same two levels available in cells where the same machinery implements both the applications (interactions) and the operating system (events).

A critical task in the translation of biological models into our language and generally into process algebras, is the specification of multi-reactant multi-product reactions. These reactions, rare in nature [21], are quite frequent in biological models (see for example [23]) as abstractions of sequences of elementary steps whose details are unknown or not of interest. Since actions in the BlenX language involve at most two processes, a possible way to translate multi-reactant multi-product reactions is to decompose them into a sequence of one-reactant or two-reactant reactions. Consider for instance the a system composed of four species, R_1, R_2, R_3 and P, that can interact through the reaction:

$$R_1 + R_2 + R_3 \xrightarrow{k} P$$

where R_1, R_2, and R_3 are the reactants, P the product and k is the reaction rate constant[1]. Furthermore, we can assume that the species R_3 can be degradeted with reaction constant rate k_d. If we have no biological information about the former reaction we can try to decompose it in the following two reactions with rates k_1 and k_2 respectively:

$$R_1 + R_2 \xrightarrow{k_1} R_1 : R_2$$
$$R_1 : R_2 + R_3 \xrightarrow{k_2} P$$

where $R_1 : R_2$ is the intermediate complex of the first two reactants. If this approach is adopted, some problems arise, as described below.

- Given n reactants, there are $\frac{n!}{2}$ possible ways to decompose a multi-reactant multi-product reaction.
- A reaction may block at intermediate steps leading to a deadlock. This may happen for instance if the reactant R_3 misses or it is consumed in the degradation reaction. If we put the first reaction reversible, it is possible to come back to the original situation, but also in this case we may model behaviours as a sequence of complexation and decomplexation that consumes resources and produce no useful outcome.
- Quantitatively, we have to assign the two constant rates k_1 and k_2 and see what is the relation between them and the rate k.

We will consider this simple example to illustrate our approach in the rest of the paper.

In order to overcome the drawbacks highlighted above we proposed the TBlenX language, an extension of BlenX with transactions.

Transactions are mechanisms originally used in web services and databases to execute distributed computations as if they were a single atomic action. Recently there have been different attempts to model web-service transactions by using process algebras [3,6,27,28,5,7,8]. These previous works will be discussed in Section 2. Here we do not consider neither compensation and rollback mechanisms nor nested transactions

---

[1] This is the constant rate used in the deterministic formulation of chemical reactions. Specifically the associated kinetics is given by the mass-action law and it is given by $k \cdot R_1 \cdot R_2 \cdot R_3$.

nor time as these features are not necessary to describe biological reactions. We focus on simple transactions that have to satisfy the properties of *atomicity* and *serializability*, suitable for modeling biology. Atomicity is summarized as "all or nothing": either a transaction is executed and finally commits or it does nothing. Serializability expresses that different activities have the same effect whether they are executed in sequence or in parallel. Hereafter these transactions are called *biological transactions.*

The paper is organized in the following way. Section 2 reports some related works. In Section 3 an introduction to the BlenX language is reported. In Section 4 the extended calculus is described in detail. Some properties are introduced in the following section. Some examples are shown in Section 6. Finally, the last section reports discussion and some final remarks.

2   Related Works

There are various works concerning the application of process algebras and formal methods for the modelling of biological systems [15,9,4,20,12]. In PEPA [22,9] it is possible to express multi-reactant multi-product reactions by means of multi synchronization along a common action. However, the level of abstraction proposed by this language is high and some biological details, such as the kind of interaction or the sites of biological elements, cannot be expressed in an explicit way. The same approach has been considered in Bio-PEPA [11,12], an extension of PEPA for the modelling and analysis of biochemical networks. BIOCHAM [20] is a programming environment for modeling biochemical systems, making simulations and querying the model in temporal logic. By using this language it is possible to represent multi-molecular complexes, the sites of the proteins and the locations (compartments) where species are. BIOCHAM formal objects represent chemical or biochemical species, ranging from small molecules to macromolecules and genes, and reaction rules represent biological interactions. Both BIOCHAM and BlenX can represent a large quantity of biological interactions and structures (for instance the interaction sites), but they are based on two different approaches: BIOCHAM evolves from rewriting rule languages whereas the BlenX evolves from process algebras. A main peculiarity of the latter with respect to the former language is the *communication* and, specifically, the definition of *affinity* between interaction sites. This feature allows us to represent some biological processes based on the compatibility for the interaction: the same substance can interact with more elements in the context, although with different levels of affinity.

Previous works concerning the application of web-service transactions to process algebras have been studied in [3,6,27,28,5,7,8]. In [3] the πt-calculus is presented: it is an extension of the asynchronous π-calculus extended to deal with long time transactions and offers failure handlers when interruptions are met. Another extension of the asynchronous π-calculus with long-time transactions, called *web-π*, is introduced in [27]. In this case the main aspects are the interruptible processes, the failure handlers and the concept of time. A web-π transaction may terminate successfully or may fail, either as an error occurs or the time deadline is reached. CSP is the process algebras adopted in [8] to model long-running transactions with traces. The authors of [7] introduced a new calculus, *StAC* (Structural Activity Compensation), inspired by both CCS and CSP, to model long-running business transactions. It gives a precise interpretation of compensation, including the combination of compensation with parallel execution, hierarchy and exceptions. Finally, in [6] a formal study of the serializability of transactions in JavaSpaces is undertaken. For this purpose the authors abstract away from the concrete language and embed the primitives in a process calculus. None of these works are in the context of systems biology. An extension of the π-calculus with transactions in the context of biological systems has been presented in [13].

3 The BlenX Language

This section recalls the syntax and the semantics of the BlenX language, as presented in [17,19,18]. It is based on the concept of *boxes*, abstracting biological entities, with some sites (*binders*) to express the interaction capabilities of the element, in which π-like processes (*processes*) are encapsulated.

A BlenX program, also called β-system, is a tuple $Z = \langle B, E, \xi \rangle$ which is a composition of a set *boxes* B, a list of *events* E and an *environment* ξ. The element B intuitively represents the structure of the system, E represents the list of possible events enabled on the system and the environment ξ contains information like the set *Types* of considered types (ranged over by $\Delta, \Gamma_0, \Sigma', \cdots$).

3.1 The Syntax

We briefly describe here the syntax of BlenX, for more details see [17,19,18]. The boxes and the list of events E are defined according to the following context-free grammar:

$$
\begin{array}{lll}
B & ::= \mathsf{Nil} \mid B[P] \mid B\|B \\
B & ::= \widehat{\beta}(x, r, \Delta) \mid \widehat{\beta}(x, r, \Delta)B \\
\widehat{\beta} & ::= \beta \mid \beta^h \mid \beta^c \\
P & ::= \mathsf{nil} \mid P|P \mid !\pi.P \mid M \qquad M ::= \pi.P \mid M + M \\
\pi & ::= x(y) \mid \overline{x}\langle y\rangle \mid (\tau, r) \mid (\mathsf{die}, r) \mid (\mathsf{ch}(x, \Delta), r) \mid \\
& \quad (\mathsf{hide}(x), r) \mid (\mathsf{unhide}(x), r) \mid (\mathsf{expose}(x, s, \Delta), r) \\
cond & ::= B[P]; r \mid |B[P]| = n \mid B[P], B[P]; r \\
verb & ::= \mathsf{new}(B[P], n) \mid \mathsf{split}(B[P], B[P]) \mid \mathsf{join}(B[P]) \mid \mathsf{delete} \\
event & ::= (cond) \; verb \\
E & ::= event \mid event :: E
\end{array}
$$

where we assume a countably infinite set $\mathcal{N}$ of names (ranged over by the lower-case letters $x, y, ...$) and $n \in \mathbb{N}$. Furthemore, the special name $\tau \notin \mathcal{N}$ expresses internal activities of processes or delays and $r \in \mathbb{R}^+ \cup \{\infty\}$ is the rate parameter. The rate ∞ denotes immediate actions whereas r stands for the *stochastic* (or *basal*) rate, for details about this see Section 3.3.

Boxes generated by the non terminal symbol B can be either *elementary* or a parallel composition of elementary boxes. The special process Nil is the deadlock. The box $B[P]$ is a process prefixed by a specialized *binder* B that represents the interaction capabilities of the box. A binder B is made up of a non empty list of *elementary binders* of the form $\beta(x, r, \Gamma)$ (*active*), $\beta^h(x, r, \Gamma)$ (*hidden*) or $\beta^c(x, r, \Gamma)$ (*complexed*) ($\widehat{\beta}$ stands for a binder in any of the previous forms), where the name x is the *subject* of the binder and acts as a binder for the free occurrences of x in the internal processes, r is a rate parameter and Γ represents the type of x. The letters $\mathcal{P}$, $\mathcal{B}$ and $\mathbf{\mathcal{B}}$ denote the set of all the possible processes,boxes and binders, respectively. The functions $sub : \mathbf{\mathcal{B}} \rightarrow 2^N$, $types : \mathbf{\mathcal{B}} \rightarrow 2^{Types}$ and $bc : \mathbf{\mathcal{B}} \rightarrow 2^{Types}$ return the subject, the types and the set of complexed elements of a box. The definition of *well-formed* boxes is reported below (the operator $| - |$ denotes the cardinality of the argument).

Definition 1. *Let* $B = \mathbf{B}_1[P_1] \parallel \cdots \parallel \mathbf{B}_n[P_n]$. *We say that B is well-formed if* $\forall i \in \{1, ..., n\}$. $\mathbf{B}_i$ *is well-formed.*
A binder is well-formed if $|\mathbf{B}_i| = |sub(\mathbf{B}_i)| = |types(\mathbf{B}_i)| > 0$.

A binder is *well-formed* if it is a non-empty string of elementary binders where subjects and types are all distinct.

Processes generated by the non terminal symbol P are referred as *processes*. In addition to the usual action proposed in *pi*-calculus (except for restriction that is removed here), there are actions peculiar to the BlenX language. The action (die, r) destroys the box enclosing the process that executes the prefix. The action $(ch(x, \Delta), r)$ changes the type of the binder with subject x to Δ. The actions $(hide(x), r)$ and $(unhide(x), r)$ are complementary and they change the state of an active binder to hidden and vice versa. Finally, the action $(expose(x, s, \Delta), r)$ creates a new binder for the current box with subject x, rate s and type Δ.

The term E generates a list of events. Each single event occurs only if its condition is satisfied on a set of one or more boxes composing B. A single *event* is the composition of a condition *cond* and an action *verb*. Events replace the f_{join} and f_{split} axioms of the original Beta-binders definition [33] for the sake of implementation.

The syntactic category *verb* denotes the actions that are associated with conditions. The action $new(\mathbf{B}[P], n)$ creates n new instances of the box specified as argument. Hereafter when the created box coincides with the one specified in the condition the notation $new(n)$ is used for short. The action $split(\mathbf{B}[P], \mathbf{B}[P])$ removes a copy of the box in the condition and introduces the two processes arguments of the split operation. The action $join(\mathbf{B}[P])$ removes a copy of each of the boxes in the condition and introduces a copy of its argument. Finally, the action delete removes a copy of the box reported in the condition.

Biologically speaking, the action new can be used to model the translation of new proteins in the cell at a given rate, the actions split and join can represent classical bind/unbind reactions in molecular environments and, finally, delete is useful to model the decay or degradation of molecules and proteins.

The definition of *well-formed* events is reported below. First we need to define the structural congruence for BlenX.

Definition 2. *The structural congruence over processes, denoted* $\equiv_p$, *is the smallest congruence relation which satisfies the laws in Fig. 1 (group a), the structural congruence over boxes, denoted* $\equiv_b$, *is the smallest congruence relation which satisfies the laws in Fig. 1 (group b) and the structural congruence over events, denoted* $\equiv_e$, *is the smallest congruence relation which satisfies the laws in Fig. 1 (group c).*

group a
1- $P_1 \equiv_p P_2$ if P_1 and P_2 α-equivalent
2- $P_1 \mid (P_2 \mid P_3) \equiv_p (P_1 \mid P_2) \mid P_3$
3- $P_1 \mid P_2 \equiv_p P_2 \mid P_1$
4- $P \mid$ nil $\equiv_p P$
5- $M_1 + (M_2 + M_3) \equiv_p (M_1 + M_2) + M_3$
6- $M_1 + M_2 \equiv_p M_2 + M_1$
7- $!\pi.P \equiv_p \pi.(P \mid !\pi.P)$

group b
8- $\boldsymbol{B}[P_1] \equiv_b \boldsymbol{B}[P_2]$ if $P_1 \equiv_p P_2$
9- $B_1 \parallel (B_2 \parallel B_3) \equiv_b (B_1 \parallel B_2) \parallel B_3$
10- $B_1 \parallel B_2 \equiv_b B_2 \parallel B_1$
11- $B \parallel$ Nil $\equiv_b B$
12- $\boldsymbol{B}_1 \boldsymbol{B}_2[P] \equiv_b \boldsymbol{B}_2 \boldsymbol{B}_1[P]$
13- $\boldsymbol{B}^* \widehat{\beta}(x, r, \Gamma)[P] \equiv_b \boldsymbol{B}^* \widehat{\beta}(y, r, \Gamma)[P\{y/x\}]$
with y fresh in P and $y \notin sub(\boldsymbol{B}^*)$

group c
14- $(B_0; r)$ $split(B_1, B_2) \equiv_e (B_0'; r)$ $split(B_1', B_2')$
if $B_0 \equiv_b B_0'$, $B_1 \equiv_b B_1'$ and $B_2 \equiv_b B_2'$
15- $(B; r)$ $delete \equiv_e (B'; r)$ $delete$, if $B \equiv_b B'$
16- $(B; r)$ $new(B_1, n) \equiv_e (B'; r)$ $new(B_1', n)$
if $B \equiv_b B'$ and $B_1 \equiv_b B_1'$
17- $(\lvert B \rvert = m)$ $new(B_1, n) \equiv_e (\lvert B' \rvert = m)$ $new(B_1', n)$
if $B \equiv_b B'$ and $B_1 \equiv_b B_1'$
18- $(B_0, B_1; r)$ $join(B_2) \equiv_e (B_0', B_1'; r)$ $join(B_2')$
if $B_0 \equiv_b B_0'$, $B_1 \equiv_b B_1'$ and $B_2 \equiv_b B_2'$
19- $E_0 :: E_1 \equiv_e E_1 :: E_0$

Fig. 1. Structural laws for BlenX

Definition 3. *Let* (cond) verb *be an event. We say that the event is well-formed if it satisfies one of the following forms and conditions:*

- $(\boldsymbol{B}_1[P_1]; r)$ split$(\boldsymbol{B}_2[P_2], \boldsymbol{B}_3[P_3])$
  with $(bc(\boldsymbol{B}_2) \cap bc(\boldsymbol{B}_3) = \emptyset)$ and $(bc(\boldsymbol{B}_2) \cup bc(\boldsymbol{B}_3) = bc(\boldsymbol{B}_1))$;
- $(\boldsymbol{B}_1[P_1], \boldsymbol{B}_2[P_2]; r)$ join$(\boldsymbol{B}_3[P_3])$
  with $((bc(\boldsymbol{B}_1) \cap bc(\boldsymbol{B}_2) = \emptyset)$ and $(bc(\boldsymbol{B}_1) \cup bc(\boldsymbol{B}_2) = bc(\boldsymbol{B}_3)))$;
- $(\boldsymbol{B}[P]; r)$ new$(\boldsymbol{B}'[P'], n)$ with $bc(\boldsymbol{B}) = \emptyset$;
- $(\lvert \boldsymbol{B}[P] \rvert = m)$ new$(\boldsymbol{B}'[P'], n)$ with $bc(\boldsymbol{B}) = \emptyset$ and $\boldsymbol{B}'[P'] \equiv_b \boldsymbol{B}[P]$;
- $(\boldsymbol{B}[P]; r)$ delete with $bc(\boldsymbol{B}) = \emptyset$.

A list E of events is well-formed if all its events are well-formed.

The intuition underlying the above definition is that manipulation of boxes must take care of complexes. In particular, it is forbidden to create new copies or to destroy boxes that are part of complexes (last three items).

The environment ξ contains the following components:

- the set *Types* of types, that determines the interaction capabilities;
- the affinity function $\alpha : Types^2 \to \mathbb{R}^3$, that determines the interaction rate;
- the function $\rho : N \to \mathbb{R}$, that returns the rates associated with channel names;
- the symmetric binary relation $\odot$, called complexation relation.

Given two types Δ and Γ, the application $\alpha(\Delta, \Gamma)$ returns (r, s, t) where the value r is the *complexation* stochastic rate, s is the *decomplexation* stochastic rate and t is the *inter-communication* stochastic rate. The auxiliary functions $\alpha_c(\Delta, \Gamma) = r$, $\alpha_d(\Delta, \Gamma) = s$ and $\alpha_i(\Delta, \Gamma) = t$ project the components of the result of $\alpha(\Delta, \Gamma)$. The function ρ associates stochastic rates to names in N and drives the stochastic behaviour of the communications enabled on the channel passed as argument. Note that the environment is defined by the user.

In order to define the complexation relation, we need to introduce the following definition:

Definition 4. Let $\mathcal{H} = \{||_0, ||_1\}$ and $\vartheta \in \mathcal{H}^*$. The set of labels H (with metavariable γ) is defined as $\gamma ::= \vartheta\Delta$, where $\Delta \in Types$.

Labels ϑ are also called localities [16] and provide a linear encoding of the syntactical location of the interaction sites in the syntax tree of the whole initial system. Indeed the tags $||_0$ and $||_1$ denote the left and right branch in the tree and Δ provides interaction sites of boxes with unique names.

Intuitively, the complexation relation $\odot \subseteq H \times H$ states that two interaction sites are joined in a complex. We say that the relation $\odot$ is well-formed if for each pair $(\gamma, \gamma') \in \odot$ there does not exist another pair in $\odot$ that contains γ or γ'.

Definition 5. Let $\odot$ be a complexation relation. The relation $\odot$ is well-formed if $\gamma \odot \gamma' \wedge \gamma \odot \gamma'' \Rightarrow \gamma' = \gamma''$.

Two labels are connected if there exists a path of relations built by $\odot$ that relates the two labels, i.e., the corresponding interaction sites are part of the same complex.

Definition 6. Let $\odot$ be a well-formed complexation relation and let $\gamma, \gamma' \in \odot$. Then γ and γ' are connected, denoted with $\gamma\overline{\odot}\gamma'$, if there exists labels $\gamma_1 = \vartheta_1\Delta_1, \cdots, \gamma_n = \vartheta_n\Delta_n$ such that

$$(\gamma = \gamma_1) \wedge (\gamma' = \gamma_n) \wedge (\forall i \in \{1, ..., n/2\}\gamma_{2i-1} \odot \gamma_{2i}) \wedge$$
$$(\forall i \in \{1, ..., (n/2) - 1\}(\vartheta_{2i} = \vartheta_{2i+1} \wedge \Delta_{2i} \neq \Delta_{2i+1}))$$

3.2   The Operational Semantics

Each box $B[P]$ is replaced with a labeled box $\vartheta B[P]$. The definition of structural congruence for the BlenX language has been reported above in Fig. 1. This congruence is decidable and efficiently solvable (see [35]). Moreover, we have:

Definition 7. *Two β-systems $Z = \langle B, E, \xi \rangle$ and $Z' = \langle B', E', \xi' \rangle$ are structurally congruent, indicated with $Z \equiv Z'$, only if $B \equiv_b B'$, $E \equiv_e E'$ and $\xi = \xi'$.*

Formally, assuming B and B', we have that:

$$\vartheta B \equiv_b \vartheta' B' \Leftrightarrow \vartheta = \vartheta' \wedge B \equiv_b B'$$

The operational semantics of the language is defined using a reduction relation $\xrightarrow{r}_s$, based on a labeled reduction relation $\xrightarrow{\theta}$.

Definition 8. *The set of labels Θ, with metavariable θ, is defined in the following way:*

$$\theta ::= r, kind, data$$

where $r \in \mathbb{R} \cup \{\infty\}$, $kind \in \{die, new\}$ and data is a generic string. The function rate : $\Theta \to \mathbb{R}$ returns the value r of the triple.

In the semantics rules, we use the symbol • for action kinds different from *die* and *new* and the symbol ϵ for the empty string. It is possible to distinguish between three types of operations: *monomolecular*, *bimolecular* and *events*. A brief description of these actions is reported in the following paragraphs.

Monomolecular operations. The formal semantics of monomolecular operations is reported in Table 1. Hereafter substitutions are typed as $\{-/-\} : \mathcal{N} \to \mathcal{N}$. The first rule describes the *intra-box communication*, i.e. how components within the same box interact. If the channel x appears in the binders the rate associated with the intra-communication is s, otherwise, if x is a free name in the box the rate associated with the communication is $\rho(x)$. The *expose* action adds a new site of interaction to the interface, the *change* action modifies the type of an interaction site, the *hide* and *unhide* actions make respectively invisible and visible an interaction site. The *die* action eliminates the box that performs the action and, by propagating the proper information with the label $\theta = r, die, \vartheta$ through the derivation tree, allows the elimination of all the boxes directly or indirectly complexed with the one performing the action. In the rule die the notation $\xi[\circ']$ is used to indicate the substitution of the complexation relation with a new one that records the modifications.

Bimolecular operations. The formal semantics of bimolecular operations is reported in Table 2. *Inter-communication* represents the notion of communication between boxes. In particular, the communication is enabled only if the affinity of the types of the involved elementary binders $\alpha(\Delta, \Gamma) = (0, 0, n)$ with $n > 0$. This means that the complexation and decomplexation feature is not enabled and hence only a notion of communication is permitted. *Complex* and *decomplex* operations create and delete dedicated communication binding between boxes. If we consider $\alpha(\Delta, \Gamma) = (r, s, t)$ with $r, s, t > 0$ we have that the *complex* operation creates a communication binding with rate $\alpha_c(\Delta, \Gamma)$ and the *decomplex* operation deletes an already existing binding with rate $\alpha_d(\Delta, \Gamma)$. Finally, the *inter-complex communication*, described by the rule (inter_c), enables a communication between complexed boxes through the complexed sites.

Table 1. Monomolecular reduction rules for BlenX

(intra_c) $\langle \vartheta B[\overline{x}\langle z\rangle. P_1 + M_1 \mid x(w). P_2 + M_2 \mid P_3], E, \xi\rangle \xrightarrow{r,\bullet,\epsilon} \langle \vartheta B[P_1 \mid P_2\{z\!/\!w\} \mid P_3], E, \xi\rangle$

where $r = s$ if $B = B_0^* \beta(x, s, \Delta) B_1^*$ while $r = \rho(x)$ otherwise

(tau) $\langle \vartheta B[(\tau, r). P_1 + M_1 \mid P_2], E, \xi\rangle \xrightarrow{r,\bullet,\epsilon} \langle \vartheta B[P_1 \mid P_2], E, \xi\rangle$

(expose) $\langle \vartheta B[(\text{expose}(x, s, \Gamma), r). R + M \mid Q], E, \xi\rangle \xrightarrow{r,\bullet,\epsilon} \langle \vartheta B \beta(y, s, \Gamma)[R\{y\!/\!x\} \mid Q], E, \xi\rangle$

$y \notin sub(B)$ and $\Gamma \notin types(B)$

(change) $\langle \vartheta B^* \beta(x, s, \Delta)[(\text{ch}(x, \Gamma), r). R + M \mid Q], E, \xi\rangle \xrightarrow{r,\bullet,\epsilon} \langle \vartheta B^* \beta(x, s, \Gamma)[R \mid Q], E, \xi[\odot']\rangle$

where $\odot' = (\odot_\xi \setminus \{(\gamma, \gamma') \in \odot_\xi : \gamma = \vartheta\Delta \vee \gamma = \vartheta\Delta\}) \cup (\odot \cup \odot^{-1})$

with $\odot = \{(\vartheta\Gamma, \gamma) \in \odot_\xi : \vartheta\Delta \odot_\xi \gamma\}$

(hide) $\langle \vartheta B^* \beta(x, s, \Delta)[(\text{hide}(x), r). R + M \mid Q], E, \xi\rangle \xrightarrow{r,\bullet,\epsilon} \langle \vartheta B^* \beta^h(x, s, \Delta)[R \mid Q], E, \xi\rangle$

(unhide) $\langle \vartheta B^* \beta^h(x, s, \Delta)[(\text{unhide}(x), r). R + M \mid Q], E, \xi\rangle \xrightarrow{r,\bullet,\epsilon} \langle \vartheta B^* \beta(x, s, \Delta)[R \mid Q], E, \xi\rangle$

(die) $\langle \vartheta B[(\text{die}, r). R + M \mid Q], E, \xi\rangle \xrightarrow{r,die,\vartheta} \langle \text{Nil}, E, \xi[\odot']\rangle$

where $\odot' = \odot_\xi \setminus \{(\vartheta_0\Delta, \vartheta_1\Gamma) : \vartheta_0 = \vartheta \vee \vartheta_1 = \vartheta\}$

Table 2. Bimolecular reduction rules for BlenX

	$P_1 \equiv {}_p\overline{x}\langle z\rangle. R_1 + M_1 \mid Q_1$	$P_2 \equiv {}_p y(w). R_2 + M_2 \mid Q_2$

(inter) $\langle \vartheta_1 B_1[P_1] \parallel \vartheta_2 B_2[P_2], E, \xi\rangle \xrightarrow{\alpha_i(\Gamma,\Delta),\bullet,\epsilon} \langle \vartheta_1 B_1[R_1 \mid Q_1] \parallel \vartheta_2 B_2[R_2\{z\!/\!w\} \mid Q_2], E, \xi\rangle$

provided $\alpha_c(\Gamma, \Delta) = 0$ and where $B_1 = \beta(x, r, \Delta) B_1^*$, $B_2 = \beta(y, s, \Gamma) B_2^*$ and $z \notin sub(B_2)$

(comp) $\langle \vartheta_1 \beta(x, r, \Delta) B_1^*[P_1] \parallel \vartheta_2 \beta(y, s, \Gamma) B_2^*[P_2], E, \xi\rangle \xrightarrow{r,\bullet,\epsilon} \langle \vartheta_1 B_1[P_1] \parallel \vartheta_2 B_2[P_2], E, \xi[\odot]\rangle$

where $B_1 = \beta^c(x, r, \Delta) B_1^*$, $B_2 = \beta^c(y, s, \Gamma) B_2^*$, $r = \alpha_c(\Gamma, \Delta)$

and $\odot = \odot_\xi \cup \{(\vartheta_1\Delta, \vartheta_2\Gamma), (\vartheta_2\Gamma, \vartheta_1\Delta)\}$

(dcomp) $\langle \vartheta_1 \beta^c(x, r, \Delta) B_1^*[P_1] \parallel \vartheta_2 \beta^c(y, s, \Gamma) B_2^*[P_2], E, \xi\rangle \xrightarrow{r,\bullet,\epsilon} \langle \vartheta_1 B_1[P_1] \parallel \vartheta_2 B_2[P_2], E, \xi[\odot]\rangle$

where $B_1 = \beta(x, r, \Delta) B_1^*$, $B_2 = \beta(y, s, \Gamma) B_2^*$, $r = \alpha_d(\Gamma, \Delta)$

and $\odot = \odot_\xi \setminus \{(\vartheta_1\Delta, \vartheta_2\Gamma), (\vartheta_2\Gamma, \vartheta_1\Delta)\}$

	$P_1 \equiv {}_p\overline{x}\langle z\rangle. R_1 + M_1 \mid Q_1$	$P_2 \equiv {}_p y(w). R_2 + M_2 \mid Q_2$

(inter_c) $\langle \vartheta_1 B_1[P_1] \parallel \vartheta_2 B_2[P_2], E, \xi\rangle \xrightarrow{\alpha_i(\Gamma,\Delta),\bullet,\epsilon} \langle \vartheta_1 B_1[R_1 \mid Q_1] \parallel \vartheta_2 B_2[R_2\{z\!/\!w\} \mid Q_2], E, \xi\rangle$

where $B_1 = \beta^c(x, r, \Delta) B_1^*$ and $B_2 = \beta^c(y, s, \Gamma) B_2^*$, $\vartheta_1\Delta \odot_\xi \vartheta_2\Gamma$ and $z \notin sub(B_2)$

Events. Events can be considered as global rules of the environment, triggered only when the conditions associated with them are satisfied.

The formal semantics of events is reported in Table 3. The meaning of the event conditions is the following:

– $(B[P]; r)$verb: The action *verb* is enabled, with rate r, only if the element B of the β-system Z is structurally congruent to $\vartheta B[P] \parallel B'$;

- $(B[P], B'[P']; r)$verb: The action *verb* is enabled, with rate r, only if B in the β-system Z is structurally congruent to $\vartheta B[P] \parallel \vartheta' B'[P'] \parallel B'$;
- $(\|B[P]\| = m)$verb: The action *verb* is enabled, with rate ∞, only B in the β-system Z is structurally congruent to the $\underbrace{\vartheta_0 B[P] \parallel \cdots \parallel \vartheta_m B[P]}_{m} \parallel B'$ and $B' \not\equiv_b \vartheta B[P] \parallel B''$.

The *verb* component can be a *split*, a *join*, a *new* and a *delete* action. The *split* action can occur in a well-formed event of the form

$$(B[P]; r) \text{ split } (B_1[P_1], B_2[P_2])$$

The meaning of this action is that, if the condition is satisfied, the split action substitutes an occurrence of a box structurally congruent to $B[P]$ in B with the parallel composition of the boxes $B_1[P_1]$ and $B_2[P_2]$, with rate r. The *join* action can occur in a well-formed event of the form

$$(B_1[P_1], B_2[P_2]; r) \text{ join } (B[P])$$

If the condition is satisfied in the β-system Z, the execution of the *join* action, enabled with rate r, substitutes an occurrence of boxes structurally congruent to $B_1[P_1]$ and $B_2[P_2]$ in B with the box $B[P]$. Both *split* and *join* actions produce modifications in the environment ξ. The *delete* action can occur in a well-formed event of the form

$$(B[P]; r) \text{ delete}$$

The execution of the *delete* action consumes one instance of a box structurally congruent to $B[P]$ in B, with rate r. Finally, the *new* action is described by the well-formed events

$$(B[P]; r) \text{ new } (B'[P'], n) \text{ and } (\|B[P]\| = m) \text{ new } (B'[P'], n)$$

These events are enabled (the first with rate r and the second with rate infinite), only if B contains at least a box for the first event and exactly m boxes for the second event that are structurally congruent to $B'[P']$. The execution of the event, in both cases, creates n copies of the box $B'[P']$.

In the present semantics the actual number of boxes $B[P]$ in the whole system cannot be derived only from the new axiom. This problem is solved by using a labeled semantics: the new axiom propagate a label $\theta = r, new, (B'[P'], m, n)$ which contains the information about the *new* action.

The stochastic transition system. Before explaining the behaviour of the last rules, some definitions are needed.

Definition 9. *Let $Z = \langle B, E, \xi \rangle$ be a β-system with $\xi = (T, \alpha, \rho, \odot)$. Then Z is well-formed only if B is well-formed, E is well-formed, $\odot$ is well-formed and*

$$(\vartheta\Delta, \vartheta'\Gamma) \in \odot \Leftrightarrow (B \equiv_b \vartheta \beta^c(x, r, \Delta) B_1^*[P_1] \parallel \vartheta' \beta^c(y, s, \Gamma) B_2^*[P_2] \parallel B')$$

Table 3. Events reduction rules for BlenX

(split)	$\langle \vartheta B[P], E, \xi \rangle \xrightarrow{r, \bullet, \epsilon} \langle \vartheta((\|_0 B_0[P_0]) \| (\|_1 B_1[P_1])), E, \xi[\odot'] \rangle$

where $E = (B[P]; r)$ split$(B_0[P_0], B_1[P_1]) :: E'$ and

$\odot' = (\odot_\xi \setminus \odot_0) \cup (\odot_1 \cup \odot_1^{-1} \cup \odot_2 \cup \odot_2^{-1})$ with

$\odot_0 = \{(\vartheta_0 \Delta_0, \vartheta_1 \Delta_1) \in \odot_\xi : \vartheta = \vartheta_0 \vee \vartheta = \vartheta_1\},$

$\odot_1 = \{(\vartheta_0 \Delta_0, \vartheta \|_0 \Delta) : \vartheta_0 \Delta_0 \odot_\xi \vartheta \Delta\},$

$\odot_2 = \{(\vartheta_1 \Delta_1, \vartheta \|_1 \Delta) : \vartheta_1 \Delta_1 \odot_\xi \vartheta \Delta\}$

(join)	$\langle \vartheta_0 B_0[P_0] \| \vartheta_1 B_1[P_1], E, \xi \rangle \xrightarrow{r, \bullet, \epsilon} \langle \vartheta_0 B[P] \| \text{Nil}, E, \xi[\odot'] \rangle$

where $E = (B_0[P_0], B_1[P_1]; r)$ join$(B[P]) :: E'$ and

$\odot' = (\odot_\xi \setminus \odot_0) \cup (\odot_1 \cup \odot_1^{-1})$ with

$\odot_0 = \{(\vartheta \Delta, \vartheta' \Delta') \in \odot_\xi : \vartheta_1 = \vartheta \vee \vartheta_1 = \vartheta'\},$

$\odot_1 = \{(\vartheta \Delta, \vartheta_0 \Delta') : \vartheta \Delta \odot_\xi \vartheta_1 \Delta'\}$

(delete)	$\langle \vartheta B[P], (B[P]; r) \text{ delete}() :: E, \xi \rangle \xrightarrow{r, \bullet, \epsilon} \langle \text{Nil}, (B[P] : r) \text{ delete} :: E, \xi \rangle$
(new)	$\langle \vartheta B[P], (|B[P]| = m) \text{ new}(B'[P'], n) :: E, \xi \rangle \xrightarrow{\theta} \langle \vartheta B[P], (|B[P]| = m) \text{ new}(B'[P'], n) :: E, \xi \rangle$

where $\theta = \infty, new, (B'[P'], m, n)$

(new_c)	$\langle \vartheta B[P], (B[P]; r) \text{ new}(B'[P'], n) :: E, \xi \rangle \xrightarrow{r, \bullet, \epsilon} \langle \vartheta B, (B[P]; r) \text{ new}(B'[P'], n) :: E, \xi \rangle$

where $B = (\|_0 B[P]) \| \underbrace{(\|_1 \|_0 B'[P']) \| (\|_1 (\cdots \| (\|_1 B'[P']))}_{n}$

It can be useful to give a formal notion of *complex*.

Definition 10. *Let $\odot$ be a well-formed complexation relation and let $\vartheta B[P]$ and $\vartheta' B'[P']$ be well-formed boxes. The box $\vartheta B[P]$ is connected with the box $\vartheta' B'[P']$, denoted with $\vartheta B[P] \overline{\odot} \vartheta' B'[P']$, if*

$$\exists \Delta \in bc(B), \Gamma \in bc(B') \text{ such that } \vartheta \Delta \overline{\odot} \vartheta' \Gamma$$

A set of boxes completely connected together can be considered a complex. Note that the notion of complex is not explicit in the language, but it is a consequence of the presence of complex and decomplex operations.

Definition 11. *Let $Z = \langle B, E, \xi \rangle$ be a well formed β-system and let $B \equiv_b B' \| B''$ where $B' = \vartheta_1 B_1[P_1] \| \cdots \| \vartheta_n B_n[P_n]$. The element B' is a complex in B only if,*

$$\forall i \in \{1, ..., n\}((\exists j \in \{1, ..., n\}(i \neq j \wedge \vartheta_i B_i[P_i] \overline{\odot} \vartheta_j B_j[P_j])) \wedge$$
$$(\nexists \vartheta B[P] \text{ in } B'' : (\vartheta_i B_i[P_i] \overline{\odot} \vartheta B[P])))$$

The last three reduction rules are reported in Table 11. The struct rule, which is standard in reduction semantics, equates the behaviours of structurally congruent β-systems.

The redex rule is used to collect the context and uses a function $\mathfrak{C} : \mathcal{B} \times \vartheta \times \odot \to \mathcal{B}$ to update complexes as a consequence of a die operation. It is defined on the structure of labeled boxes in the following way:

$$\mathfrak{C}(\vartheta' B[P], \vartheta, \odot) = \begin{cases} (\text{Nil}, \odot') & \text{if } \exists \Delta, \Gamma \in T : \vartheta \Delta \overline{\odot} \vartheta' \Gamma \text{ and} \\ & \odot' = \{(\vartheta_0 \Delta, \vartheta_1 \Gamma) \in \odot : \vartheta_0 = \vartheta' \vee \vartheta_1 = \vartheta'\} \\ (\vartheta' B[P], \odot) & \text{otherwise} \end{cases}$$

$$\mathfrak{C}(\vartheta' B[P] \parallel B, \vartheta, \odot) = \mathfrak{C}(\vartheta' B[P], \vartheta, \odot) @ \mathfrak{C}(B, \vartheta, \odot)$$

with the function @ defined as:

$$(\vartheta B, \odot) @ (\vartheta' B', \odot') = (\vartheta B \parallel \vartheta' B', \odot \cup \odot').$$

The redex_s rule is used for constructing the actual transition relation. We introduce this additional level of derivation because of the presence of a particular type of *new* event. This rule uses the function Num (Table 5) that counts the number of boxes structurally congruent to $B[P]$ that are present in B.

A final reduction rule performs the check for the global condition and represents the transition relation of the stochastic reduction system. Formally,

Definition 12. *The* BlenX *Stochastic Transition System (STS) is referred as*
$S = (\mathcal{Z}, \xrightarrow{r}_s, \mathcal{Z}_0)$, *where* $\mathcal{Z}$ *is the set of well-formed β-systems,* $\mathcal{Z}_0 \in \mathcal{Z}$ *is the initial β-system and* $\xrightarrow{}_s \subseteq \mathcal{Z} \times \mathbb{R} \times \mathcal{Z}$ *is the stochastic reduction relation, where r is a stochastic rate.*

Table 4. The reduction rules for BlenX

(struct)	$\dfrac{Z_1 \equiv Z_1' \quad Z_1 \xrightarrow{\theta} Z_2}{Z_1' \xrightarrow{\theta} Z_2}$
(redex)	$\dfrac{\langle B, E, \xi \rangle \xrightarrow{\theta} \langle B', E, \xi' \rangle}{\langle B \parallel B_1, E, \xi \rangle \xrightarrow{\theta} \langle B' \parallel B_2, E, \xi'[\odot'] \rangle}$
	where $(B_2, \odot) = \mathfrak{C}(B_1, \vartheta, \odot_\xi)$ and $\odot' = \odot_{\xi'} \setminus \odot$ if $\theta = (r, die, \vartheta)$, while $B_2 = B_1$ and $\odot' = \odot_{\xi'}$ otherwise
(redex_s)	$\dfrac{\langle B, E, \xi \rangle \xrightarrow{\theta} \langle B', E, \xi' \rangle}{\langle B, E, \xi \rangle \xrightarrow{r}_s \langle B' \parallel B_1, E, \xi' \rangle}$
	where $B_1 = \overbrace{(\parallel_0 B[P]) \parallel (\parallel_1 (\cdots \parallel (\parallel_1 B[P]))}^{n}$ if $\theta = (r, new, (B[P], m, n))$ and $\text{Num}(B[P], B) = m$, while $B_1 = \text{Nil}$ otherwise

Table 5. Auxiliary function Num used in the stochastic reduction relation

$\text{Num}(B[P], \text{Nil}) = 0$
$\text{Num}(B[P], B'[P'] \parallel B) = 1 + \text{Num}(B[P], B)$ if $B[P] \equiv_b B'[P']$
$\text{Num}(B[P], B'[P'] \parallel B) = \text{Num}(B[P], B)$ if $B[P] \not\equiv_b B'[P']$

The definition of *STS* is built upon the set of well-formed β-systems $\mathcal{Z}$. Finally note that the structural congruence and $\xrightarrow{r}_s$ reduction preserve the well-formedness of β-systems (see [39].

3.3   The Stochastic Algorithm

BlenX refers to the stochastic Gillespie's algorithm [21] for the simulation. Specifically, the algorithm proposed is an efficient variant of the original algorithm. The interested reader can refer to [39] for the details.

The Gillespie's algorithm is a widely-used method for the simulation of biochemical reactions. It finds an *exact* solution[2] of the *Chemical Master Equation (CME)*, that describes the transition of a biological system from one state to another through changes of the probability of the system being in a certain state. Finding a solution of this equation directly is unfeasible (with very few exceptions).

This algorithm deals with homogenous, well-stirred systems in thermal equilibrium and constant volume, composed of N different species that interact through M reactions. Broadly speaking, the goal is to describe, starting from an initial state, the evolution of the system $\mathbf{X}(t) = (X_1(t), X_2(t), ..., X_N(t))$, where $X_i(t)$ stands for the number of molecules of the species i. Note that the original version of the algorithm refers to *elementary reactions*[3].

Every reaction is characterized by a stochastic rate constant c_j, called *basal rate*. This is derived from the reaction constant rate k by means of some simple relations proposed in [21,40]. With the basal rate, it is possible to calculate the *actual rate* of the reaction, that is the probability of R_j happening in time $(t, t + \Delta t)$ given the system in a specific state. The actual rate for the reaction R_j is calculated as

$$a_j = c_j \cdot h(\mathbf{X}(t))$$

where $h(\mathbf{X}(t))$ is a function that calculates the number of possible combinations of reactants in the system. In the case of a monomolecular reaction $A \rightarrow ...$ the number of possible combination is simply $|A|$ (where $|\cdot|$ gives the amount of A), whereas in the case of the bimolecular reaction $A + B \rightarrow ...$ we have that the number of combination of the reactants is $|A| \cdot |B|$, if A and B are different, or $(|A| \cdot |A - 1|)/2$ if they are the same species.

The algorithm is based essentially on the following two steps:

– calculation of the next reaction that occurs in the system;
– calculation of the time when the next reaction occurs.

We derive the information above from two conditional density functions: $p(j|\mathbf{X}(t)) = a_j(\mathbf{X}(t))/a_0$, that is the probability that the next reaction is R_j and $p(\tau|\mathbf{X}(t)) = a_0 e^{a_0 \mathbf{X}(t)\tau}$, the probability that the next reaction occurs in $[t + \tau, t + \tau + d\tau]$, where $a_0 = \sum_{v=1}^{M} a_v(\mathbf{X}(t))$.

---

[2] In the sense that the algorithm produces possible time-evolution trajectories that are consistent with CME governing the physical process.

[3] Elementary reactions are interactions involving at most two reactants. So they include zeroth-order interactions, monomolecular and bimolecular reactions. In the context of reaction rate equations their dynamics is described by the mass-action law and a constant rate.

We observe that from the semantics rules we can infer the basal rate r associated with the transition and from this the actual rate. Given a β-system, the derivation of the actual rate associated with the transition is obtained by deriving the number of boxes in the system that are congruent to the boxes involved in the interaction. This information is kept implicit in the model and then calculated at the moment of the simulation.

As observed above, the original Gillespie's algorithm refers to elementary reactions and this is the case that is considered in the current release of the simulator tool [39] because the BlenX language considers only elementary interactions. However the extension to multi-reactant multi-product reactions is widely-used in different simulation tools based on Gillespie's method and is presented formally in [40].

3.4 Example: Three-Reactant One-Product Reaction

Consider the biological system introduced in Section 1. It is composed of four species and two reactions.

In BlenX each species corresponds to a box. We have that the species R_1, R_2, R_3 and P are represented by the boxes $B_1 = \beta(x_1, \infty, \Gamma_1)B_1[\text{nil}]$, $B_2 = \beta(x_2, \infty, \Gamma_2)B_2[\text{nil}]$, $B_3 = \beta(x_3, \infty, \Gamma_3)B_3[\text{nil}]$ and $B_P = B_P[\text{nil}]$, respectively.

As discussed before, the three-reactant one-product reaction must be decomposed into two join events: the former represents the merge of the first two reactants R_1 and R_2 and the latter the merge of the resulting intermediate complex $R_1 : R_2$ and R_3. These events are:

$$(\boldsymbol{B}_{R_1}, \boldsymbol{B}_{R_2}; r_1) \ \text{join} \ (\boldsymbol{B}_{R_1:R_2});$$

$$(\boldsymbol{B}_{R_1:R_2}, \boldsymbol{B}_{R_3}; r_2) \ \text{join} \ (\boldsymbol{B}_P).$$

The intermediate complex is represented by the box $\boldsymbol{B}_{R_1:R_2}[\text{nil}]$, with $\boldsymbol{B}_{R_1:R_2} = \beta(x_{R_1:R_2}, \infty, \Gamma_{R_1:R_2})$. The rates r_1 and r_2 are the basal rates associated with the two elementary reactions and here are assumed known.

The degradation of the species R_3 with basal rate r_d is defined in BlenX by means of a delete event:

$$(\boldsymbol{B}_{R_3}; r_d) \ \text{delete}$$

A possible reduction of the system $S = B_1 \parallel B_2 \parallel B_3 \parallel S'$ is:

$$B_1 \parallel B_2 \parallel B_3 \parallel S' \xrightarrow{r_1}_s \boldsymbol{B}_{R_1:R_2}[\text{nil}] \parallel B_3 \parallel S'$$
$$\xrightarrow{r_d}_s \boldsymbol{B}_{R_1:R_2}[\text{nil}] \parallel \text{Nil} \parallel S'$$

The former reaction has started but it stops at an intermediate step as the third reactant has been consumed in the second transition.

4 The BlenX Language with Biological Transactions

In this section we report the extension to the syntax and the semantics of the language needed to implement biological transactions. We call this extension the TBlenX language.

4.1 Biological Transactions

Transactions are the basic mechanism for modeling database transactions and for composing web-services in orchestration and choreography languages. Different properties and features must be considered according to the field of application. In the literature transactions generally have some complex features such as compensation processes or nesting or timeout mechanisms [3,6,27,28,5,7,8].

Modeling biological interactions require transactions that satisfy simpler properties. In particular, transactions should end successfully as the sequence of actions were atomic (*atomicity*). Furthermore, reaction results should be visible only after transactions have ended (*serializability*).

We refer to the transactions described here as *"biological transactions"*, to distinguish them from database and web service ones. These transactions are considered in [13], where both the π-calculus and the biochemical stochastic π-calculus are enriched with biological transactions.

4.2 General Ideas

Our extension is based on the following ideas:

- the transaction names t, t', t'', ...$\in \mathcal{T}$ are introduced to identify transactions. These names are added to the syntax of the BlenX language to indicate in which transactions the boxes and the processes are involved. Given a transaction t, a box can be either *blocked* (it is part of) or *unblocked* (it is not part of) with respect to t.
- An action characterized by a set of names $\mathcal{T}$ can be executed only if the respective box is blocked by $t \in \mathcal{T}$. If a box is unblocked, only the actions with $T = \emptyset$ can be executed.
- We need to introduce two new events to represent the *start* and the *end* of the transaction. The event start describes the block of the boxes involved in a transaction $t \in \mathcal{T}$. The event *end* describes the unblock of the final box and the consequent end of the transaction.
- The standard reduction relation for BlenX is extended in order to consider transactions. Two new axioms start_t and end_t are added.
- We define a TBlenX program, called also βT-system as a tuple $Z = \langle B, E, \xi \rangle$, where the components are modified with respect to the original definition in order to take into account transactions. It is worth noting that we use the same names to indicate the system and the components in both BlenX and its extended version: the context will disambiguate the names.

4.3 Syntax

The syntax of the TBlenX language is defined by the following grammar.

$$B \quad ::= \mathsf{Nil} \mid (\boldsymbol{B}[P])^{t^*} \mid B\|B$$

$$\boldsymbol{B} \quad ::= \widehat{\beta}(x,r,\varDelta) \mid \widehat{\beta}(x,r,\varDelta)\boldsymbol{B}$$

$$\widehat{\beta} \quad ::= \beta \mid \beta^h \mid \beta^c$$

$$P \quad ::= \mathsf{nil} \mid P|P \mid !\pi^{T^*}.P \mid M \qquad M ::= \pi^{T^*}.P \mid M + M$$

$$\pi \quad ::= x(y) \mid \overline{x}\langle y \rangle \mid (\tau, r) \mid (\mathsf{die}, r) \mid (\mathsf{ch}(x,\varDelta), r) \mid$$
$$\qquad (\mathsf{hide}(x), r) \mid (\mathsf{unhide}(x), r) \mid (\mathsf{expose}(x, s, \varDelta), r)$$

$$v \quad ::= [] \mid [n \diamond (\boldsymbol{B}[P])^{t^*}] :: v$$

$$cond \quad ::= v \mid v; r \mid v : t; r$$

$$verb \quad ::= \mathsf{new}(v) \mid \mathsf{split}(v) \mid \mathsf{join}(v) \mid \mathsf{delete} \mid \mathsf{start}(v) \mid \mathsf{end}(v)$$

$$event ::= (cond) \; verb$$

$$E \quad ::= event \mid event :: E$$

We assume a countable set of transaction names $\mathcal{T}$ (ranged over by lower-case letters t, t', t'', ...), with $\mathcal{N} \cap \mathcal{T} = \emptyset$. We use t^* to denote a transaction name in $\mathcal{T}$ or the null string (denoted by ϵ). Furthermore, T denotes a non-empty subset of transaction names in $\mathcal{T}$. The set T^* may be either T or the empty set. If $T^* = \emptyset$ we can omit it and we have the usual actions. With $\mathcal{PT}$, $\mathcal{BT}$ and $\overrightarrow{\mathcal{BT}}$ we denote the set of all the possible processes, boxes and binders for TBlenX, respectively.

Elementary boxes are enriched with the name t^*. If $t^* = \epsilon$, we have an *unblocked* box (corresponding to the standard box in BlenX), otherwise we have a box *blocked* by $t \in \mathcal{T}$.

In the case of processes, we add the set T^* to the prefixes for identifying the transactions in which the associated prefix/action can be involved. It is worth noting that the transactions are associated with the prefixes, not with the channel names. As a consequence, a channel name could be used in different actions involved in different transactions.

We add two events describing the *start* and the *end* of the transaction. Furthermore we modify the term *cond* to take the case of more than two boxes into account.

In order to consider stoichiometric coefficients (i.e. number of molecules of a species involved in a reaction), we follow the idea introduced in [32] and we consider the definition of a *multi box* $n \diamond B$, where $n \in \mathbb{N}$ is the *multiplicity/stoichiometry* of the box B. The multi box stands for the parallel composition of n boxes congruent to B. A list v of multi boxes is introduced in the syntax in order to collect the boxes involved in the event. The notation for v, even though more complex than the previous list of boxes, is useful to collect the information about the stoichiometry of the elements involved in the reaction and to have a more compact definition in the case of the events *start* and *end*. Note that the number of processes involved in the events *join*, *split*, *new* and *delete* is as usual, we extend the use of v to all of them to have a homogeneous description for events. The conditions on the number of boxes and on the list v is reported in the definition of well-formdeness (see below).

The list v has the form "$[n_1 \diamond B_1, n_2 \diamond B_2, ..., n_m \diamond B_m]$", with $m \geq 1$. The symbol :: stands for the concatenation of elements in the list. The list v is used both in the condition and in the verb of events. In the case of conditions, we have three possible cases. First of all, the expression v means than the condition is satisfied if we have n_1 boxes congruent to B_1, n_2 processes congruent to B_2 and so on. The action associated to the event is immediate. The expression $v; r$ has a similar meaning, but in this case the rate is r. Finally, the term $v : t; r$ is used in the events that describe the start and the end of a transaction t. The definition of v and r is as before. Note that the condition $|B[P]| = m$ for the event *new* in the BlenX language is replaced with $[m \diamond (B[P])^{t^*}]$, with the same identical meaning. If v is in the verb of the event, we have that n_1 boxes congruent to B_1, n_2 processes congruent to B_2, ... replace (or are added to, in the case of the event *new*) the processes in the condition.

Now we need to give the definition of structural congruence over processes, over boxes and over events. Here we only report the different rules with respect to Table 1.

Definition 13. *The structural congruence over processes, denoted $\equiv_p$, is the smallest congruence relation which satisfies the laws in Fig. 1 (group a, with the exception of the law 7) and the laws in Fig. 2 (group a), the structural congruence over boxes, denoted $\equiv_b$, is the smallest congruence relation which satisfies the laws in Fig. 1 (group b, with the exception of the laws 8, 12, 13) and the laws in Fig. 2 (group b) and the structural congruence over events, denoted $\equiv_e$, is the smallest congruence relation which satisfies the laws in Fig. 2 (group c).*

The law 20 claims that the process $\pi^0.P$ is congruent to the process $\pi.P$. A similar law (21) is defined for boxes as well. The law 22 concerns the congruence for the list v of boxes. The other laws are adaptations of the previous laws in the case of transactions. In the last group (c), the first six laws are similar to the ones for BlenX. The last two rules report the congruence definitions for the events describing the start and the end of a transaction.

The definition of a $\diamond$-standard form based on multi boxes follows. It is used in the semantics rules to express the fact that the action can be fired only if the boxes of the list v in the event conditions are congruent to the boxes involved in the action.

Definition 14. B *is in $\diamond$-standard form if it is* Nil *or* $B \equiv_b \prod_{i=1}^{n} n_i \diamond B_i$, *with $n \geq 1$ and*

- $n_i \geq 1$, $\forall i = 1, ..., n$;
- $B_1, ..., B_n$ *are elementary boxes*;
- $B_i \not\equiv_b B_j$, $\forall i \neq j$.

In the Definition 14, $\prod_{i=1}^{n} B_i$ is the parallel composition of n boxes and the symbol $\equiv_b$ stands for the structural equivalence for boxes in the TBlenX language as defined above. We have the following result:

Proposition 1. *Every B is structural congruent to an element B' in a $\diamond$-standard form.*

Proof. The proof is by structural induction over boxes. Three cases are possible.

- Case $B = $ Nil. B is in a $\diamond$-standard form by definition and hence $B = B'$.
- Case $B = (B[P])^{t^*}$. In this case $B \equiv_b 1 \diamond (B[P])^{t^*} = B'$ that is in $\diamond$- standard form.
- Case $B = B_1 \parallel B_2$. By inductive hypothesis on B_1 and B_2 we have that $B_1 \equiv_b \prod_{i=1}^{n_1} k_i \diamond B_{1i}$ and $B_2 \equiv_b \prod_{j=1}^{n_2} l_j \diamond B_{2j}$ with B_{1i} and B_{2j} elementary boxes for

group a
$7'$- $!\pi^{T^*}.P \equiv_p \pi^{T^*}.(P \mid !\pi^{T^*}.P)$ 20- $\pi^0.P \equiv_p \pi.P$

group b
$8'$- $(B[P_1])^{t_1^*} \equiv_b (B[P_2])^{t_2^*}$ if $P_1 \equiv_p P_2$ and $t_1^* = t_2^*$ $12'$- $(B_1 B_2[P])^{t_1^*} \equiv_b (B_2 B_1[P])^{t_2^*}$ if $t_1^* = t_2^*$ $13'$- $(B^* \widehat{\beta}(x,r,\Gamma)[P])^{t^*} \equiv_b (B^* \widehat{\beta}(y,r,\Gamma)[P\{y/x\}])^{t^*}$ with y fresh in P and $y \notin sub(B^*)$ 21- $(B)^\epsilon \equiv_b B$ 22- $v \equiv_b v'$ if v, v' are well-formed, $|v| = |v'|$ and $\forall(n_i \diamond (B_i[P_i])^{t^*})$ in v there exists $(n_j \diamond (B'_j[P'_j])^{t^*})$ in v' such that $(B_i[P_i])^{t^*} \equiv_b (B'_j[P'_j])^{t^*}$

group c
$14'$- $(v_0; r)\ split(v_1) \equiv_e (v'_0; r)\ split(v'_1)$ if $v_0 \equiv_b v'_0, v_1 \equiv_b v'_1$ $15'$- $(v; r)\ delete \equiv_e (v'; r)\ delete$, if $v \equiv_b v'$ $16'$- $(v_0; r)\ new(v_1) \equiv_e (v'_0; r)\ new(v'_1)$ if $v_0 \equiv_b v'_0$ and $v_1 \equiv_b v'_1$ $17'$- $(v_0)\ new(v_1) \equiv_e (v'_0)\ new(v'_1)$ if $v_0 \equiv_b v'_0$ and $v_1 \equiv_b v'_1$ $18'$- $(v_0; r)\ join(v_1) \equiv_e (v'_0; r)\ join(v'_1)$ if $v_0 \equiv_b v'_0$ and $v_1 \equiv_b v'_1$ $19'$- $E_0 {::} E_1 \equiv_e E_1 {::} E_0$ 23- $(v_1 : t; r)\ start(v'_1) \equiv_e (v_2 : t; r)\ start(v'_2)$ if $v_1 \equiv_b v_2$ and $v'_1 \equiv_b v'_2$ 24- $(v_1 : t; r)\ end(v'_1) \equiv_e (v_2 : t; r)\ end(v'_2)$ if $v_1 \equiv_b v_2$ and $v'_1 \equiv_b v'_2$

Fig. 2. Structural laws for TBlenX

any i and j. For any B_{1i} such that $\exists j.B_{1i} \equiv_b B_{2j}$, remove B_{2j} and the corresponding index from B_2 and replace $k_i \diamond B_{1i}$ with $(k_i + l_j) \diamond B_{1i}$ in B_1. Repeat the procedure above as long as replacements are possible. The result are two new boxes B'_1 and B'_2 such that $B'_1 \parallel B'_2 = B'$.

In order to express that a box is blocked or unblocked by a transaction we need to define the auxiliary function Act, that finds the active transactions. Similarly, it is possible to define the function act, that returns the set of transaction names that are in the prefixes of the sub-terms of a given process P. The definition of these two functions is reported in Table 6 (the symbol $\cup$ stands for the usual union of sets).

In addition to the usual components, the environment ξ contains the set of transactions $\mathcal{T}$.

Now we report the definitions of well-formdness for the language. First of all, we consider the definition of well-formdeness for the list v.

Definition 15. *A list of boxes $v = [n_1 \diamond B_1, n_2 \diamond B_2, ..., n_m \diamond B_m]$ is well-formed if the following conditions hold:*

- $|v| = m \geq 1$;
- either $\mathsf{Act}(v) = \emptyset$ or $\mathsf{Act}(n_i \diamond B_i) = \{t\}$ for some $t \in \mathcal{T}$ and $\forall i = 1, ..., m$;
- $B_i \not\equiv_b B_j$, $\forall i \neq j$.

Similarly to BlenX, we can say that a TBlenX program $\langle B, E, \xi \rangle$ is *well-formed* if all its components are well-formed. The definition of well-formdness for all the components is unchanged with the exception of events.

Definition 16. *Let* (cond) *verb be an event. We say that the event is well-formed if it satisfies one of the following forms and conditions:*

-$(v; r)$ split(v')
where either $(v = [1 \diamond (B_1[P_1])^{t_1^*}]$ *and* $v' = [1 \diamond (B_2[P_2])^{t_2^*}, 1 \diamond (B_3[P_3])^{t_3^*}]$
with $(bc(B_2) \cap bc(B_3) = \emptyset)$ *and* $(bc(B_2) \cup bc(B_3) = bc(B_1))$ *and* $t_1^* = t_2^* = t_3^*)$ *or*
$(v = [1 \diamond (B_1[P_1])^{t_1^*}]$ *and* $v' = [2 \diamond (B_2[P_2])^{t_2^*}]$ *with* $(bc(B_2) = bc(B_1))$ *and*
$t_1^* = t_2^*)$;
- $(v; r)$ join(v')
with either $(v = [1 \diamond (B_1[P_1])^{t_1^*}; 1 \diamond (B_2[P_2])^{t_2^*}]$ *and* $v' = [1 \diamond (B_3[P_3])^{t_3^*}]$ *with*
$((bc(B_1) \cap bc(B_2) = \emptyset)$ *and* $(bc(B_1) \cup bc(B_2) = bc(B_3))$ *and* $t_1^* = t_2^* = t_3^*)$
or $(v = [2 \diamond (B_1[P_1])^{t_1^*}]$ *and* $v' = [1 \diamond (B_3[P_3])^{t_3^*}]$ *with* $(bc(B_1) = bc(B_3))$ *and*
$t_1^* = t_3^*)$
- $(v; r)$ new(v') *where*
$v = [1 \diamond (B[P])^{t^*}]$ *with* $bc(B) = \emptyset$ *and* $v' = [m \diamond (B'[P'])^{t^*}]$;
- (v) new(v') *where* $v = [n \diamond (B[P])^{t^*}]$ *with* $bc(B) = \emptyset$
and $v' = [m \diamond (B'[P'])^{t^*}]$;
- (v) delete *where* $v = [1 \diamond (B[P])^{t^*}]$ *with* $bc(B) = \emptyset$.
- $(v : t; r)$ start(v')
with $\mathsf{Act}(v) = \emptyset$ *and* $\mathsf{Act}(v') = \{t\}$,
$v = [\kappa_1 \diamond (B_1[P_1]), \kappa_2 \diamond (B_2[P_2]), ..., \kappa_{n_r} \diamond (B_{n_r}[P_{n_r}])]$, *where* $n_r \geq 1$ *and*
$v' = [\kappa_1 \diamond (B_1[P_1])^t, \kappa_2 \diamond (B_2[P_2])^t, ..., \kappa_{n_r} \diamond (B_{n_r}[P_{n_r}])^t]$
- $(v : t; r)$ end(v')
with $v = [\kappa_1 \diamond (B_1[P_1])^t, \kappa_2 \diamond (B_2[P_2])^t, ..., \kappa_{n_p} \diamond (B_{n_p}[P_{n_p}])^t]$ *where* $n_p \geq 1$
and $v' = [\kappa_1 \diamond (B_1[P_1]), \kappa_2 \diamond (B_2[P_2]), ..., \kappa_{n_p} \diamond (B_{n_p}[P_{n_p}])]$
A list E *of events is well-formed if all its events are well-formed.*

Table 6. Auxiliary functions Act and act

$\mathsf{Act}(\mathsf{Nil}) = \emptyset$
$\mathsf{Act}(B_1 \parallel B_2) = \mathsf{Act}(B_1) \cup \mathsf{Act}(B_2)$
$\mathsf{Act}((B[P])^{t^*}) = \emptyset$ if $t^* = \epsilon, \{t^*\}$ otherwise
$\mathsf{Act}(n \diamond B) = \mathsf{Act}(B)$
$\mathsf{Act}(v) = \cup_{i=1}^{m} \mathsf{Act}(n_i \diamond B_i)$ with $v = [n_1 \diamond B_1, ..., n_m \diamond B_m]$
$\mathsf{act}(\mathsf{nil}) = \emptyset$
$\mathsf{act}((\pi^{T^*}.P) = \emptyset$ if $T^* = \emptyset, T^*$ otherwise
$\mathsf{act}(P_1 \vert P_2) = \mathsf{act}(P_1) \cup \mathsf{act}(P_2)$
$\mathsf{act}(!P) = \mathsf{act}(P)$
$\mathsf{act}(M_1 + M_2) = \mathsf{act}(M_1) \cup \mathsf{act}(M_2)$

The first five forms describe the case of *join*, *split*, *new* and *delete* events. The definitions are reformulated by considering the list v and by adding the conditions over transactions. We add conditions over the list to make the multiplicity of boxes involved correspond to the events in BlenX. Specifically, a well-formed split must have one box in the condition and two in the verb, whereas the join must have two boxes in the condition and one in the verb. The event delete must have only one box in the condition, whereas the event new can have either one box (with rate r) or m boxes (in this case the rate must be ∞). As far as the transactions are concerned, it is requested that all the boxes involved in an event are either unblocked or blocked by the same transaction. The last two rules describe the events *start* and *end*. The event *start* requires the boxes in the condition being unblocked. Viceversa, the event *end* requires the boxes in the condition being blocked by a transaction.

Notation. Hereafter we use the following notation:

- $(B_1[P_1] \| B_2[P_2]... \| B_n[P_n])^{t^*}$ stands for $(B_1[P_1])^{t^*} \| (B_2[P_2])^{t^*} ... \| (B_n[P_n])^{t^*}$
- $(B)^t$ denotes one box or a set of boxes whose sub-terms are all blocked by t (and by no other transaction).
- $1 \diamond B[P]$ can be written as $B[P]$.

Example. A graphical representation of an example of a box blocked by t_1 in the TBlenXlanguage is:

$$(x : \Gamma) \qquad (y : \Delta) \qquad\qquad\qquad t_1$$

$$(\text{hide}(x), r)^{\{t_1\}}. P | (\text{unhide}(y), r)^{\{t_2\}}. Q | y(w). R$$

Only the first process, $(\text{hide}(x), r)^{\{t_1\}}. P$, is active as the prefix is blocked by t_1. The process $(\text{unhide}(y), r)^{\{t_2\}}. Q$ may be executed only when the box is blocked by t_2 (the box is labelled by t_2) and finally $y(w). R$ refers to the case in which the box is unblocked.

4.4 Semantics

The operational reduction semantics is based on a structural congruence and a reduction relation. The *structural congruence* has been defined in Fig. 2. We have the following definition for congruence in the case of βT-systems.

Definition 17. *Two βT-systems $Z = \langle B, E, \xi \rangle$ and $Z' = \langle B', E', \xi' \rangle$ are structurally congruent, written $Z \equiv Z'$, only if $B \equiv_b B'$, $E \equiv_e E'$ and $\xi = \xi'$.*

We extend the second component of the label θ in the BlenX language. This modification is necessary to study the properties of the transactions.

Definition 18. *The set of transition labels Φ (ranged over by $\phi_1, \phi_2, ...$) is defined as:*

$$\phi = r, l, data$$

where r and data are as defined for the label θ in the reduction system for the **BlenX**
*language and l ∈ L describes the action type. The set L (ranged over by l_1, l_2, ...) is
defined as:*

$$\mathcal{L}_1 \cup \mathcal{L}_2 \cup \mathcal{L}_3 \cup \{t : start, t : end\} \cup \{t : \alpha | \alpha \in \mathcal{L}_1 \cup \mathcal{L}_2 \cup \mathcal{L}_3\}$$

with:

- $\mathcal{L}_1 = \{i, tau, e, c, h, u, die\}$ *is the set of molecular actions;*
- $\mathcal{L}_2 = \{I, Ic, C, D\}$ *is the set of bimolecular actions;*
- $\mathcal{L}_3 = \{S, J, D, newc, new\}$ *is the set of the event actions different from start and
  end;*
- *The label t : β (where β is end, start or any other action in $\mathcal{L}_1 \cup \mathcal{L}_2 \cup \mathcal{L}_3$) indicates
  that the transition of kind β regards boxes blocked by t.*

The following definition introduces the *stochastic transition system* for **TBlenX**.

Definition 19. *The* **TBlenX** *Stochastic Transition System (TSTS) is referred as* $S = (\mathcal{Z}, \overset{\phi}{\Rightarrow}, Z_0)$, *where* $\mathcal{Z}$ *is the set of well-formed βT-systems,* $Z_0 \in \mathcal{Z}$ *is the initial βT-
system and* $\overset{\phi}{\Rightarrow} \subseteq \mathcal{Z} \times \mathbb{R} \times \mathcal{Z}$ *is the smallest relation over boxes obtained by applying
the axioms and rules in Tables 8, 9, 10 and 11.*

The main differences with the reduction rules proposed in Section 3 are the addition of
the names $t \in \mathcal{T}$, the set T of transaction names, the axioms start_t and end_t. The rate
associated to the transition is handled as in **BlenX**.

Table 8 reports the reduction rules for *monomolecular reactions*. The rule intra_t
represents the communication inside a box. A condition about transactions is added: a
communication along a channel x is possible only if the box is blocked by $t^* \in T_1^* \cap T_2^*$
or $t^* = \epsilon$ and $T_1^* = T_2^* = \emptyset$, where T_1^* and T_2^* are the set of transaction names in which
the input and the output prefixes may be involved. All the other rules in Table 8 have the
set T^* associated to the prefix and the action is possible if the respective box is blocked
by a transaction $t^* \in T^*$ or ($t^* = \epsilon$ and $T^* = \emptyset$).

Table 9 reports the rules for *bimolecular reactions*, modified to consider transactions.
The rules inter_t and inter_c_t are enriched by the condition that a communication is
possible only if the boxes are either blocked by $t^* \in T_1^* \cap T_2^*$ or ($t^* = \epsilon$ and $T_1^* = T_2^* = \emptyset$),
where T_1^* and T_2^* are the set of transaction names in which the input and the output
prefixes may be involved. Similar conditions are added to the other rules in this table.

Table 10 reports the rules for *events*. The boxes involved in the events must be either
unblocked or all blocked by the same transaction. The first five rules are adaptations of
the ones presented for **BlenX** to the case of transactions. The last two rules concern the
start and the *end* of the transaction. The rule start_t describes the start of a transaction t
by allowing the block of the boxes involved in the transaction via t. The side condition
"Act(B) = ∅" guarantees that the initial box is unblocked. The rule end_t is used to
define what happens when a transaction t ends successfully: the boxes blocked by t are
unblocked.

The last two rules are reported in Table 11. The modifications concern the rule re-
dex_t. The idea is to add some side conditions concerning the transactions active in
the system to guarantee that the actions involved in the transactions have the priority

Table 7. Auxiliary function Num used in the reduction relation for TBlenX

$$\text{Num}((\boldsymbol{B}[P])^{t^*}, \text{Nil}) = 0$$
$$\text{Num}((\boldsymbol{B}[P])^{t^*_1}, (\boldsymbol{B}'[P'])^{t^*_2}\|B) = 1 + \text{Num}((\boldsymbol{B}[P])^{t^*_1}, B) \quad \text{if } (\boldsymbol{B}[P])^{t^*_1} \equiv_b (\boldsymbol{B}'[P'])^{t^*_2}$$
$$\text{Num}((\boldsymbol{B}[P])^{t^*_1}, (\boldsymbol{B}'[P'])^{t^*_2}\|B) = \text{Num}((\boldsymbol{B}[P])^{t^*_1}, B) \quad \text{if } (\boldsymbol{B}[P])^{t^*_1} \not\equiv_b (\boldsymbol{B}'[P'])^{t^*_2}$$

over the other actions. In the case of the start of a transaction no other transactions must be currently executed in the system. This is obtained by considering the condition $\text{Act}(B_1) = \emptyset$. In the case the action is involved in a transaction, we have the condition that $\text{Act}(B_1) \subseteq \text{Act}(B)$, i.e. the rest of the system is blocked by the same transaction of B or it is unblocked. The other conditions reported in the cases of the *die* and the *new* actions are the same as those of BlenX. The auxiliary function Num has been modified in order to consider transactions. The definition is reported in Table 7.

4.5 Rates

Following [17], we consider Gillespie's method as the reference algorithm, but other ones could be considered as well. Since transactions represent reactions with more than two reactants, an extended version of Gillespie's approach is necessary [40].

Table 8. Monomolecular reduction rules for TBlenX

(intra_t) $\quad \langle \vartheta(\boldsymbol{B}[(\overline{x}\langle z\rangle)^{T^*_1}\langle z\rangle.P_1 + M_1 \mid (x(w))^{T^*_2}.P_2 + M_2 \mid P_3])^{t^*}, E, \xi\rangle \xrightarrow{r,l,\epsilon} \langle \vartheta(\boldsymbol{B}[P_1 \mid P_2\{z/w\} \mid P_3])^{t^*}, E, \xi\rangle$

where $r = s$ if $\boldsymbol{B} = \boldsymbol{B}^*_0 \beta(x, s, \varDelta)\boldsymbol{B}^*_1$ while $r = \rho(x)$ otherwise, and

either $(t^* \in T^*_1 \cap T^*_2$ and $l = t : i$ with $t^* = t)$ or $(l = i$ and $t^* = \epsilon$ and $T^*_1 = T^*_2 = \emptyset)$

(tau_t) $\quad \langle \vartheta(\boldsymbol{B}[(\tau, r)^{T^*}.P_1 + M_1 \mid P_2])^{t^*}, E, \xi\rangle \xrightarrow{r,l,\epsilon} \langle \vartheta(\boldsymbol{B}[P_1 \mid P_2])^{t^*}, E, \xi\rangle$

provided that either $(t^* \in T^*$ and $l = t : tau$ with $t^* = t)$ or $(t^* = \epsilon$ and $T = \emptyset$ and $l = tau)$

(expose_t) $\quad \langle \vartheta(\boldsymbol{B}[((\text{expose}(x, s, \Gamma), r))^{T^*}.R + M \mid Q])^{t^*}, E, \xi\rangle \xrightarrow{r,l,\epsilon} \langle \vartheta(\boldsymbol{B}\,\beta(y, s, \Gamma)[R\{y/x\} \mid Q])^{t^*}, E, \xi\rangle$

where $y \notin sub(\boldsymbol{B})$ and $\Gamma \notin types(\boldsymbol{B})$

and provided that either $(t = t^* \in T^*$ and $l = t : e$ with $t = t^*)$ or $(l = e$ and $t^* = \epsilon$ and $T = \emptyset)$

(change_t) $\quad \langle \vartheta(\boldsymbol{B}^*\,\beta(x, s, \varDelta)[(\text{ch}(x, \Gamma), r)^{T^*}.R + M \mid Q])^{t^*}, E, \xi\rangle \xrightarrow{r,l,\epsilon} \langle \vartheta(\boldsymbol{B}^*\,\beta(x, s, \Gamma)[R \mid Q])^{t^*}, E, \xi\rangle$

where $\odot' = (\odot_\xi \setminus \{(\gamma, \gamma') \in \odot_\xi : \gamma = \vartheta\varDelta \vee \gamma = \vartheta\varDelta\}) \cup (\odot \cup \odot^{-1})$

provided that either $(t^* \in T^*$ and $l = t : c$ with $t = t^*)$ or $(t^* = \epsilon$ and $T = \emptyset$ and $l = c)$

(hide_t) $\quad \langle \vartheta(\boldsymbol{B}^*\,\beta(x, s, \varDelta)[((\text{hide}(x), r))^{T^*}.R + M \mid Q])^{t^*}, E, \xi\rangle \xrightarrow{r,l,\epsilon} \langle \vartheta(\boldsymbol{B}^*\,\beta^h(x, s, \varDelta)[R \mid Q])^{t^*}, E, \xi\rangle$

provided that either $(t^* \in T^*$ and $l = t : h$ with $t = t^*)$ or $(t^* = \epsilon$ and $T = \emptyset$ and $l = h)$

(unhide_t) $\quad \langle \vartheta(\boldsymbol{B}^*\,\beta^h(x, s, \varDelta)[(\text{unhide}(x), r))^{T^*}.R + M \mid Q])^{t^*}, E, \xi\rangle \xrightarrow{r,l,\epsilon} \langle \vartheta(\boldsymbol{B}^*\,\beta(x, s, \varDelta)[R \mid Q])^{t^*}, E, \xi\rangle$

provided that either $(t^* \in T^*$ and $l = t : u$ with $t = t^*)$ or $(t^* = \epsilon, T = \emptyset$ and $l = u)$

(die_t) $\quad \langle \vartheta(\boldsymbol{B}[(die, r)^{T^*}.R + M \mid Q])^{t^*}, E, \xi\rangle \xrightarrow{r,l,\vartheta} \langle \text{Nil}, E, \xi[\odot']\rangle$

where $\odot' = \odot_\xi \setminus \{(\vartheta_0\varDelta, \vartheta_1\Gamma) : \vartheta_0 = \vartheta \vee \vartheta_1 = \vartheta\}$, with either $(t^* \in T^*$ and $l = t : die$ with $t = t^*)$ or $(t^* = \epsilon$ and $T = \emptyset$ and $l = die)$

We define the rates for each action in such a way that the actual rate associated with the start action represents the actual rate of the whole reaction. Therefore, we assume that:

- the global actual rate of the reaction is associated with the start prefix of the transaction;
- ∞ is assigned to all the other prefixes in the processes in the transaction.

It is worth noting that the simulation algorithm for BlenX must be extended in order to consider transactions. Since actions internal to a transaction are immediate, it is possible to neglect them and to simplify the algorithm. These aspects are not considered here as out of the scope of this work.

4.6   Observation

Here we comment on the relation between the BlenX language and its extension with transactions. We recover the standard formulation of the language by letting $T^* = \emptyset$ and $t^* = \epsilon$. This follows from the structural congruence in TBlenX and by noting that any TBlenX rule/axiom (with the exception of start_t and end_t not present in BlenX) when $T^* = \emptyset$ and $t^* = \epsilon$, describes the same behaviour than the respective rule/axiom in the standard calculus. The side conditions describing the transactions that are active reported in the rule redex_t may be neglected, as they are always satisfied. Finally, the axioms start_t and end_t are not considered, as they are applied only when a transaction t is introduced. As a consequence, the TBlenX language without transactions coincides with the BlenX language. If we use transactions, the transitions following the start action

Table 9. Bimolecular reduction rules for TBlenX

$$P_1 \equiv {}_p(\overline{x}\langle z\rangle)^{T_1^*}\langle z\rangle.R_1 + M_1 \mid Q_1 \qquad P_2 \equiv {}_p(y(w))^{T_2^*}.R_2 + M_2 \mid Q_2$$

(inter_t) $\langle\vartheta_1(\boldsymbol{B}_1[P_1])^{t^*} \parallel \vartheta_2(\boldsymbol{B}_2[P_2])^{t^*}, E, \xi\rangle \xRightarrow{\alpha_i(\Delta,\Gamma),l,\epsilon} \langle\vartheta_1(\boldsymbol{B}_1[R_1 \mid Q_1])^{t^*} \parallel \vartheta_2(\boldsymbol{B}_2[R_2\{z\!/w\} \mid Q_2])^{t^*}, E, \xi\rangle$

where $\alpha_c(\Gamma,\Delta) = 0, \alpha_i(\Gamma,\Delta) > 0$, $\boldsymbol{B}_1 = \beta(x,r,\Gamma)\boldsymbol{B}_1^*, \boldsymbol{B}_2 = \beta(y,s,\Delta)\boldsymbol{B}_2^*$ and $z \notin sub(\boldsymbol{B}_2)$ and
provided that either ($t^* \in T^*_1 \cap T^*_2$ and $l = t : I$ with $t = t^*$) or ($t^* = \epsilon, T^*_1 = T^*_2 = \emptyset$ and $l = I$)

(comp_t) $\langle\vartheta_1(\beta(x,r,\Delta)\boldsymbol{B}_1^*[P_1])^{t_1^*} \parallel \vartheta_2(\beta(y,s,\Gamma)\boldsymbol{B}_2^*[P_2])^{t_2^*}, E, \xi\rangle \xRightarrow{\alpha_c(\Delta,\Gamma),l,\epsilon} \langle\vartheta_1(\boldsymbol{B}_1[P_1])^{t_1^*} \parallel \vartheta_2(\boldsymbol{B}_2[P_2])^{t_2^*}, E, \xi[\odot]\rangle$

where $\boldsymbol{B}_1 = \beta^c(x,r,\Delta)\boldsymbol{B}_1^*, \boldsymbol{B}_2 = \beta^c(y,s,\Gamma)\boldsymbol{B}_2^*$ and $\odot = \odot_\xi \cup \{(\vartheta_1\Delta,\vartheta_2\Gamma),(\vartheta_2\Gamma,\vartheta_1\Delta)\}$
$\alpha_c(\Delta,\Gamma) > 0$, and provided that either $t_1^* = t_2^*$ and ($l = t : C$ or $l = C$)

(dcomp_t) $\langle\vartheta_1(\beta^c(x,r,\Delta)\boldsymbol{B}_1^*[P_1])^{t_1^*} \parallel \vartheta_2(\beta^c(y,s,\Gamma)\boldsymbol{B}_2^*[P_2])^{t_2^*}, E, \xi\rangle \xRightarrow{\alpha_d(\Delta,\Gamma),l,\epsilon} \langle\vartheta_1(\boldsymbol{B}_1[P_1])^{t_1^*} \parallel \vartheta_2(\boldsymbol{B}_2[P_2])^{t_2^*}, E, \xi[\odot]\rangle$

where $\boldsymbol{B}_1 = \beta(x,r,\Delta)\boldsymbol{B}_1^*, \boldsymbol{B}_2 = \beta(y,s,\Gamma)\boldsymbol{B}_2^*$ and $\odot = \odot_\xi \setminus \{(\vartheta_1\Delta,\vartheta_2\Gamma),(\vartheta_2\Gamma,\vartheta_1\Delta)\}$,
$\alpha_d(\Delta,\Gamma) > 0$ and provided that either $t_1^* = t_2^*$ and ($l = t : D$ or $l = D$)

$$P_1 \equiv {}_p(\overline{x}\langle z\rangle)^{T_1^*}\langle z\rangle.R_1 + M_1 \mid Q_1 \qquad P_2 \equiv {}_p(y(w))^{T_2^*}.R_2 + M_2 \mid Q_2$$

(inter_c_t) $\langle\vartheta_1(\boldsymbol{B}_1[P_1])^{t^*} \parallel \vartheta_2(\boldsymbol{B}_2[P_2])^{t^*}, E, \xi\rangle \xRightarrow{\alpha_i(\Delta,\Gamma),l,\epsilon} \langle\vartheta_1(\boldsymbol{B}_1[R_1 \mid Q_1])^{t^*} \parallel \vartheta_2(\boldsymbol{B}_2[R_2\{z\!/w\} \mid Q_2])^{t^*}, E, \xi\rangle$

where $\boldsymbol{B}_1 = \beta(x,r,\Delta)\boldsymbol{B}_1^*, \boldsymbol{B}_2 = \beta(y,s,\Gamma)\boldsymbol{B}_2^*, \vartheta_1\Delta \odot_\xi \vartheta_2\Gamma, \alpha_i(\Gamma,\Delta) > 0$ and $z \notin sub(\boldsymbol{B}_2)$
and provided that either ($t^* \in T^*_1 \cap T^*_2$ and $l = t : Ic$ with $t^* = t$) or
($t^* = \epsilon$ and $T^*_1 = T^*_2 = \emptyset$ and $l = Ic$)

Table 10. Events reduction rules for the TBlenX language

(split_t) $\langle \vartheta(\boldsymbol{B}[P])^{t^*}, E, \xi \rangle \overset{r,l,\epsilon}{\Longrightarrow} \langle \vartheta((\|_0 \ (\boldsymbol{B}_0[P_0])^{t^*}) \ \| \ (\|_1 \ (\boldsymbol{B}_1[P_1])^{t^*})), E, \xi[\circleddash'] \rangle$

     where $E = (v; r) \ \mathsf{split}(v') :: E'$, $v = [1 \ \diamond \ (\boldsymbol{B}[P])^{t^*}]$,

     either $(v' = [1 \ \diamond \ (\boldsymbol{B}_0[P_0])^{t^*}; 1 \ \diamond \ (\boldsymbol{B}_1[P_1])^{t^*}])$ or

     $(v' = [2 \ \diamond \ (\boldsymbol{B}_0[P_0])^{t^*}]$ with $(\boldsymbol{B}_0[P_0])^{t^*} \equiv_b (\boldsymbol{B}_1[P_1])^{t^*})$ and

     $\circleddash' = (\circleddash_\xi \setminus \circleddash_0) \cup (\circleddash_1 \cup \circleddash_1^{-1} \cup \circleddash_2 \cup \circleddash_2^{-1})$ with

     $\circleddash_0 = \{(\vartheta_0 \Delta_0, \vartheta_1 \Delta_1) \in \circleddash_\xi : \vartheta = \vartheta_0 \vee \vartheta = \vartheta_1\}$,

     $\circleddash_1 = \{(\vartheta_0 \Delta_0, \vartheta \ \|_0 \ \Delta) : \Delta \in types(\boldsymbol{B}_0) \wedge \vartheta_0 \Delta_0 \ \circleddash_\xi \ \vartheta \Delta\}$,

     $\circleddash_2 = \{(\vartheta_1 \Delta_1, \vartheta \ \|_1 \ \Delta) : \Delta \in types(\boldsymbol{B}_1) \wedge \vartheta_1 \Delta_1 \ \circleddash_\xi \ \vartheta \Delta\}$

(join_t) $\langle \vartheta_0 (\boldsymbol{B}_0[P_0])^{t^*} \ \| \ \vartheta_1 (\boldsymbol{B}_1[P_1])^{t^*}, E, \xi \rangle \overset{r,l,\epsilon}{\Longrightarrow} \langle \vartheta_0 (\boldsymbol{B}[P])^{t^*} \rangle \ \| \ \mathsf{Nil}, E, \xi[\circleddash'] \rangle$

     where $E = (v; r) \ \mathsf{join}(v') :: E'$,

     either $(v = [1 \ \diamond \ (\boldsymbol{B}_0[P_0])^{t^*}; 1 \ \diamond \ (\boldsymbol{B}_1[P_1])^{t^*}])$ or

     $(v = [2 \ \diamond \ (\boldsymbol{B}_0[P_0])^{t^*}]$ with $(\boldsymbol{B}_0[P_0])^{t^*} \equiv_b (\boldsymbol{B}_1[P_1])^{t^*})$ and $v' = [1 \ \diamond \ (\boldsymbol{B}[P])^{t^*}]$ and

     $\circleddash' = (\circleddash_\xi \setminus \circleddash_0) \cup (\circleddash_1 \cup \circleddash_1^{-1})$ with

     $\circleddash_0 = \{(\vartheta \Delta, \vartheta' \Delta') \in \circleddash_\xi : \vartheta_1 = \vartheta \vee \vartheta_1 = \vartheta'\}$,

     $\circleddash_1 = \{(\vartheta \Delta, \vartheta_0 \ \|_0 \ \Delta') : \Delta' \in types(\boldsymbol{B}_1) \wedge \vartheta \Delta \ \circleddash_\xi \ \vartheta_1 \Delta'\}$

(delete_t) $\langle \vartheta(\boldsymbol{B}[P])^{t^*}, (v;r) \ \mathsf{delete} :: E, \xi \rangle \overset{r,l,\epsilon}{\Longrightarrow} \langle \mathsf{Nil}, (v;r) \ \mathsf{delete} :: E, \xi \rangle$

     where $v = [1 \ \diamond \ (\boldsymbol{B}[P])^{t^*}]$

(new_t) $\langle \vartheta(\boldsymbol{B}[P])^{t^*}, (v) \ \mathsf{new}(v') :: E, \xi \rangle \overset{\phi}{\Longrightarrow} \langle \vartheta \boldsymbol{B}[P], (v) \ \mathsf{new}(v') :: E, \xi \rangle$

     where $v = [m \ \diamond \ (\boldsymbol{B}[P])^{t^*}]$ and $v' = [n \ \diamond \ (\boldsymbol{B}'[P'])^{t^*}]$,

     either $(\phi = (new, \infty, (\boldsymbol{B}'[P'], m, n))$ and $t^* = \epsilon)$ or $(\phi = (t : new, \infty, ((\boldsymbol{B}'[P'])^t, m, n))$ and $t^* = t)$

(new_c_t) $\langle \vartheta(\boldsymbol{B}[P])^{t^*}, ((v; r)) \ \mathsf{new}((v')) :: E, \xi \rangle \overset{r,l,\epsilon}{\Longrightarrow} \langle \vartheta \boldsymbol{B}, ((v; r)) \ \mathsf{new}((v')) :: E, \xi \rangle$

     where $v = [1 \ \diamond \ (\boldsymbol{B}[P])^{t^*}]$ and $v' = [n \ \diamond \ (\boldsymbol{B}'[P'])^{t^*}]$ and

     $B = (\|_0 \ (\boldsymbol{B}[P]))^{t^*} \ \| \ \underbrace{(\|_1 \ (\cdots \| \ (\|_1 \ (\boldsymbol{B}'[P']))^{t^*})}_{n}$

(start_t) $\langle \vartheta B, E, \xi \rangle \overset{r,t:start,\epsilon}{\Longrightarrow} \langle \vartheta(B)^t, E, \xi[\circleddash'] \rangle$

     with $Act(B) = \emptyset$ and $E = (v; t : r) \ \mathsf{start}(v') :: E'$ where

     $v = [\kappa_1 \ \diamond \ B_1[P_1], ..., \kappa_n \ \diamond \ B_n[P_n]]$ and $v' = [\kappa_1 \ \diamond \ (B_1[P_1])^t, ..., \kappa_n \ \diamond \ (B_n[P_n])^t]$ and

     $B \equiv_b \prod_{i=1}^{n} \kappa_i \ \diamond \ B_i[P_i]$ with $n \geq 1$

(end_t) $\langle \vartheta(B)^t, E, \xi \rangle \overset{r,t:end,\epsilon}{\Longrightarrow} \langle \vartheta B, E, \xi[\circleddash'] \rangle$

     with $E = (v : t; r) \ \mathsf{end}(v') :: E'$ where

     $v = [\kappa_1 \ \diamond \ (B_1[P_1])^t, ..., \kappa_n \ \diamond \ (B_n[P_n])^t]$ and $v' = [\kappa_1 \ \diamond \ B_1[P_1], ..., \kappa_n \ \diamond \ B_n[P_n]]$ and

     $B \equiv_b \prod_{i=1}^{n} \kappa_i \ \diamond \ B_i[P_i]$ with $n \geq 1$

---

are only internal and have the precedence over all the other external actions. They are executed one after the other and may not be interleaved with other actions. In general transactions limit the interleaving of concurrent transitions.

Table 11. The reduction rules for the TBlenX language

$$(\text{struct\_t}) \quad \frac{Z_1 \equiv Z_1' \quad Z_1 \overset{\phi}{\Rightarrow} Z_2}{Z_1' \overset{\phi}{\Rightarrow} Z_2}$$

$$(\text{redex\_t}) \quad \frac{\langle B, E, \xi \rangle \overset{\phi}{\Rightarrow} \langle B', E, \xi' \rangle}{\langle B \parallel B_1, E, \xi \rangle \overset{\phi}{\Rightarrow} \langle B' \parallel B_2, E, \xi'[\odot'] \rangle}$$

where we have the following cases:

if $\phi = (\infty, new, (\boldsymbol{B}'[P'], m, n))$ and $\text{Num}(\boldsymbol{B}[P], B) = m$, then $B_2 = B_1 \parallel B_3$ where

$$B_3 = \overbrace{(|\!|_0 \ \boldsymbol{B}[P]) \parallel (|\!|_1 \ (\cdots \parallel (|\!|_1 \ \boldsymbol{B}[P]))}^{n}) \text{ and } \text{Act}(B_1) = \text{Act}(B_2) = \text{Act}(B) = \emptyset$$

if $\phi = (\infty, t : new, ((\boldsymbol{B}'[P']^t, m, n))$ and $\text{Num}(\boldsymbol{B}[P], B) = m$, then $B_2 = B_1 \parallel B_3$ where

$$B_3 = \overbrace{(|\!|_0 \ (\boldsymbol{B}[P])^t) \parallel (|\!|_1 \ (\cdots \parallel (|\!|_1 \ (\boldsymbol{B}[P])^t))}^{n}),$$

$\text{Act}(B_1) \subseteq \text{Act}(B)$ and $\text{Act}(B_3) = \text{Act}(B) = \{t\}$

if $f_{action}(\phi) = die$ then $(B_2, \odot) = \mathfrak{C}(B_1, \vartheta, \odot_\xi)$ and $\odot' = \odot_{\xi'} \cap (\odot_\xi \setminus \odot)$ and

$\text{Act}(B_1) = \text{Act}(B_2) = \text{Act}(B) = \emptyset$

if $f_{action}(\phi) = t : die$ then $(B_2, \odot) = \mathfrak{C}(B_1, \vartheta, \odot_\xi)$ and $\odot' = \odot_{\xi'} \cap (\odot_\xi \setminus \odot)$ and $\text{Act}(B_1) \subseteq \text{Act}(B)$

if $(f_{action}(\phi) = \alpha$ and $\alpha \neq new, die)$ then $B_1 = B_2, \odot' = \odot_{\xi'}$ and $\text{Act}(B_1) = \text{Act}(B) = \emptyset$

if $f_{action}(\phi) = t : start$ then $B_1 = B_2, \odot' = \odot_{\xi'}$ and $\text{Act}(B_1) = \text{Act}(B) = \emptyset$

else $B_1 = B_2, \odot' = \odot_{\xi'}$ and $\text{Act}(B_1) \subseteq \text{Act}(B)$

4.7   Example: System with a Three-Reactant One-Product Reaction (Continued)

Consider the example presented in Section 1. In Section 3.4 it was shown how to translate this simple system into BlenX. Furthermore, we illustrated some drawbacks of this translation. In the following we show how this system can be modelled into TBlenX. The three-reactant one-product reaction is translated by considering two join events with the addition of the events start and end. The start and the end of the transaction are described by the following events:

$$([1 \diamond B_1, 1 \diamond B_2, 1 \diamond B_3] : t_R; r) \ \text{ start } \ ([1 \diamond (B_1)^{t_R}, 1 \diamond (B_2)^{t_R}, 1 \diamond (B_3)^{t_R}])$$

$$([1 \diamond (B_P)^{t_R}] : t_R; \infty) \ \text{ end } \ ([1 \diamond B_P])$$

where $B_1 = \beta(x_1, \infty, \Gamma_1)\boldsymbol{B}_1[P_1]$, $B_2 = \beta(x_2, \infty, \Gamma_2)\boldsymbol{B}_2[P_2]$, $B_3 = \beta(x_3, \infty, \Gamma_3)\boldsymbol{B}_3[P_3]$ and $B_P = \boldsymbol{B}_P[P_P]$ are the boxes representing the three reactants and the product. Their definition is as in BlenX. The rate r is the basal rate corresponding to the reaction constant k^4.

Two join events are defined to model the reaction and they both refer to elements blocked by the transaction t_R. The first join event represents the formation of the element $R_1 : R_2$ composed of the first two reactants R_1 and R_2. The complex is represented

---

[4] Following the definition given in [21,40] the basal rate r is derived as $k/(Na \cdot V)^2$, where Na is the Avogadro's number, i.e. the number of molecules in a mole of substance, and V is the volume of the compartment where species are. The number 2 is given by $3 - 1$ where 3 is the total number of reactants.

by the box $B_{R_1:R_2}[P_{R_1:R_2}]$, where $B_{R_1:R_2} = \beta(x_{R_1:R_2}, \infty, \Gamma_{R_1:R_2})$. The second join event represents the merge between the third reactant R_3 and the $R_1 : R_2$. The result is the box representing the product of the reaction. These events are respectively:

$$([1 \diamond (B_{R_1})^{\{t_R\}}, 1 \diamond (B_{R_2})^{\{t_R\}}]; \infty) \text{ join } ([1 \diamond (B_{R_1:R_2})^{\{t_R\}}]);$$

$$([1 \diamond (B_{R_1:R_2})^{\{t_R\}}, 1 \diamond (B_{R_3})^{\{t_R\}}]; \infty) \text{ join } ([1 \diamond (B_P)^{\{t_R\}}]).$$

The degradation reaction is translated as before.

A possible reduction of the system $S = B_1 \parallel B_2 \parallel B_3 \parallel S'$, with $\mathsf{Act}(S') = \emptyset$, is:

$$B_1 \parallel B_2 \parallel B_3 \parallel S' \xrightarrow{r,t_R:start,\epsilon} (B_1 \parallel B_2 \parallel B_3)^{t_R} \parallel S'$$
$$\xrightarrow{\infty,t_R:J,\epsilon} (R_1 : R_2 \parallel R_3)^{t_R} \parallel S'$$
$$\xrightarrow{\infty,t_R:J,\epsilon} (B_P)^{t_R} \parallel S'$$
$$\xrightarrow{\infty,t_R:end,\epsilon} B_P \parallel S'$$

Note that in this case the degradation of the species R_3 cannot happen after the start of the transaction. The complex reaction is completed.

4.8 Some Definitions

In the following we report some auxiliary definitions.

- We define two functions to extract some information from the label ϕ. The function $rate : \Phi \rightarrow \mathbb{R}^+$ returns the first component of the transition label (i.e. the rate) whereas the function $f_{action} : \Phi \rightarrow \mathcal{L}$ extracts the second component (i.e. the action type associated with the transition).
- We may distinguish two kinds of actions with respect to a transaction t:
  1. *internal transition w.r.t. t*, whose first component of the transition label is $t : \alpha$ or $t : end$;
  2. *external transition w.r.t. t*, whose first component of the transition label is $t : start$ or α.
- A function f_t is introduced to return the possible transaction in which a transition is involved. It is defined as:

$$f_t(\phi) = \begin{cases} \{t\} & \text{if } f_{action}(\phi) = t : \beta \\ \emptyset & \text{otherwise} \end{cases}$$

- $B_0 \overset{\tilde{\phi}}{\Rightarrow} B_n$ denotes the transition sequence $B_0 \overset{\phi_1}{\Rightarrow} B_1 \overset{\phi_2}{\Rightarrow} B_2 ... \overset{\phi_n}{\Rightarrow} B_n$, with $\tilde{\phi} = \phi_1 \phi_2, ... \phi_n$. The length of the transition sequence characterized by $\tilde{\phi}$ is defined as:

$$|\tilde{\phi}| = \begin{cases} 1 & \text{if } \tilde{\phi} = \phi \\ n & \text{if } \tilde{\phi} = \phi_1 \phi_2 .... \phi_n \end{cases}$$

We can now define *well-formed* transactions. In the context of biological reactions we can focus on simple transactions and suppose that after starting they end with success and errors never occur. This allows us to avoid the definition of abort actions and of an abort axiom. Furthermore no compensation mechanisms are necessary. The stop of a transaction at intermediate steps could happen when it is not possible to execute some of the actions that lead to the final boxes. As a consequence, the associate *end* event cannot be applied. We use the term *well-formed* to identify transactions that when they start, then they end successfully.

Definition 20. *A transaction t is well-formed in a βT-system $Z = \langle B, E, \xi \rangle$ if and only if the following conditions hold:*

- $t \in \mathcal{T}$;
- *if* $B_i \xRightarrow{r,t:start,\epsilon} B_1$ *with* $Act(B_i) = \emptyset$ *then there exist* $\tilde{\phi}$, B_2 *and* B_f *such that*
  1. $|\tilde{\phi}| < \infty$ *and* $f_t(\phi_i) = \{t\}$ *for each* ϕ_i *in* $\tilde{\phi}$;
  2. $Act(B_f) = \emptyset$;
  3. $B_i \xRightarrow{r,t:start,\epsilon} B_1 \xRightarrow{\tilde{\phi}} B_2 \xRightarrow{\infty,t:end,\epsilon} B_f$.
- *For each complex in B, either all its components or none of them are present in B_i.*

It is worth noting that according to the definition above a well-formed transaction always terminates with an *end* event. The possible case that the transaction ends in different ways (for instance with a *die* action) is not allowed.

Hereafter, we consider only well-formed transactions and well-formed βT-systems.

5 Properties

In this section we report some properties of the TBlenX language. Specifically, we focus on the *atomicity* and *serializability* properties. These two properties guarantee that our transactions are appropriate for describing biological reactions.

The definitions of *serialized and serializable transition sequences* proposed in [6] are modified in order to consider *bio-processes*:

Definition 21. *The transition sequence $B \xRightarrow{\tilde{\phi}} B'$, with $Act(B) = Act(B') = \emptyset$, is serialized iff $f_{action}(\phi_i) = t : \alpha$ or $f_{action}(\phi_i) = t : start$ implies $f_{action}(\phi_{i+1}) = t : \alpha$ or $f_{action}(\phi_{i+1}) = t : end$ for $i = 1, ...(n-1)$. The transition sequence $B \xRightarrow{\tilde{\phi}} B'$ is serializable, if there exists a permutation $\tilde{\phi}'$ of $\tilde{\phi}$ such that $B \xRightarrow{\tilde{\phi}'} B'$ is serialized.*

A final definition concerns the *finite derivative* of a bio-process.

Definition 22. *Given a bio-process B, a finite derivative of B is either B itself or any bio-process B' obtained by a finite transition sequence $B \xRightarrow{\tilde{\phi}} B'$.*

Now some properties are reported.

Proposition 2. *Given the bio-process B with $Act(B) = \emptyset$, for any finite derivative B' of B,*

- *there is at most one transaction active in B';*
- *there are no nested transactions in B'.*

Proof. The proof of each item follows.

- We have to show that $\mathsf{Act}(B') = \emptyset$ or $\mathsf{Act}(B') = \{t\}$ for a given $t \in \mathcal{T}$. By hypothesis, B' is a finite derivative of B and then, by definition, either $B' = B$ or there exists $\tilde{\phi}$ s.t. $B \overset{\tilde{\phi}}{\Rightarrow} B'$.

  If $B' = B$ then $\mathsf{Act}(B') = \mathsf{Act}(B) = \emptyset$ and the first item is satisfied.

  If $B' \neq B$ we prove it by induction on the length of the derivative transition sequence. Let $\tilde{\phi}$ be $\phi_1 \phi_2, ...\phi_n$ and $B \overset{\phi_1}{\Rightarrow} B_1 \overset{\phi_2}{\Rightarrow} ... \overset{\phi_{(n-1)}}{\Longrightarrow} B_{(n-1)} \overset{\phi_n}{\Rightarrow} B'$. There are two cases:

  - $length(\tilde{\phi}) = 1$. In this case $B \overset{\phi_1}{\Rightarrow} B'$ and $f_{action}(\phi_1)$ is either α or $t : start$, for a given t. In the former case $\mathsf{Act}(B') = \emptyset$ and in the latter $\mathsf{Act}(B') = \{t\}$.
  - $lenght(\tilde{\phi}) = n > 1$. By inductive hypothesis on the sequence $\tilde{\phi} = \phi_1 \phi_2, ...\phi_{(n-1)}$, we have that either $\mathsf{Act}(B_{(n-1)}) = \emptyset$ or $\mathsf{Act}(B_{(n-1)}) = \{t\}$. In the former case $f_{action}(\phi_{(n)})$ may be either $t : start$ or α, in the latter case $f_{action}(\phi_{(n)})$ may be either $t : \alpha$ or $t : end$. For any combination it is either $\mathsf{Act}(B') = \emptyset$ or $\mathsf{Act}(B') = \{t\}$.

- It sufficies to note that no bio-processes can have sub-terms of the form $(B)^t$ with $\mathsf{Act}(B) \neq \emptyset$ because it is not possible to block an already blocked bio-process. Indeed the axiom start_t can be applied only if the bio-process is unblocked.

$\square$

Proposition 3 reports some properties that immediately follow from the previous definitions.

Proposition 3.

Consider $B \overset{\phi}{\Rightarrow} B'$ with $\mathsf{Act}(B) = \{t\}$. If $f_{action}(\phi) = t : end$ then $\mathsf{Act}(B') = \emptyset$ otherwise if $f_{action}(\phi) = t : \alpha$ then $\mathsf{Act}(B') = \{t\}$.

- *If $B \overset{\phi}{\Rightarrow} B'$ with $\mathsf{Act}(B) = \emptyset$ and $f_{action}(\phi) = t : start$ then $\mathsf{Act}(B') = \{t\}$.*
- *If $B \overset{\phi}{\Rightarrow} B'$ with $\mathsf{Act}(B) = \emptyset$ and $f_t(\phi) = \emptyset$ then $\mathsf{Act}(B') = \emptyset$.*
- *Given a bio-process B such that $\mathsf{Act}(B) = \emptyset$ and given the transaction sequence $B \overset{\phi_1}{\Rightarrow} B_1 \overset{\phi_2}{\Rightarrow} B_2 ... \overset{\phi_n}{\Rightarrow} B_n$ with $f_{action}(\phi_1) = t : start$, $f_{action}(\phi_n) = t : end$ and $f_{action}(\phi_i) \neq t : end$ for all $i = 2, ..., (n-1)$ then $f_{action}(\phi_i) = t : \alpha$ for $i = 2, ...(n-1)$, $\mathsf{Act}_t(B_i) = \{t\}$ for $i = 1, ...(n-1)$ and $\mathsf{Act}(B_n) = \emptyset$.*

The next theorems concern some properties of biological transactions.

Theorem 1. *Consider the bio-process $(B_1)^t$, where t is a well-formed transaction and $\mathsf{Act}(B_1) = \emptyset$. Let B_n be the bio-process obtained by the finite transition sequence $(B_1)^t \overset{\tilde{\phi}}{\Rightarrow} B_n$ where $f_t(\phi_i) = \{t\}$ for each $\phi_i \in \tilde{\phi}$ and $\mathsf{Act}(B_n) = \emptyset$. Then $(B_1)^t \parallel S \overset{\tilde{\phi}}{\Rightarrow} B_n \parallel S$, for each S with $\mathsf{Act}(S) = \emptyset$.*

Proof. Let consider $(B_1)^t \parallel S$. By hypothesis $\mathsf{Act}(S) = \emptyset$ and $\mathsf{Act}((B_1)^t) = \{t\}$. Then, from the rule redex_t, the actions involving $(B_1)^t$ have the precedence over all the other actions. In particular, as all the actions in $\tilde{\phi}$ are involved in the transaction t and S is unblocked, the actions in $\tilde{\phi}$ have the precedence over all the others. As a consequence, $(B_1)^t \overset{\tilde{\phi}}{\Rightarrow} B_n$ implies that $(B_1)^t \parallel S \overset{\tilde{\phi}}{\Rightarrow} B_n \parallel S$. $\qquad\square$

Theorem 2. *(Atomicity) Consider the bio-process B_1, with $\mathsf{Act}(B_1) = \emptyset$. If there exists a well-defined transaction t such that $(B_1) \overset{\phi_1}{\Rightarrow} B_2 \equiv {}_b (B_1)^t$ with $f_{action}(\phi_1) = t : start$ then there exists a finite transition sequence $B_1 \overset{\phi_1}{\Rightarrow} B_2 \overset{\phi_2}{\Rightarrow} B_3 \overset{\phi_3}{\Rightarrow} \cdots \overset{\phi_{(n-1)}}{\Longrightarrow} B_n$, with $f_{action}(\phi_n) = t : end$ and $f_t(\phi_i) = \{t\}$ for each $i = 1, ..., (n-1)$.*

Proof. The result follows directly from the definition of well-defined transactions and from the operational semantics. $\qquad\square$

Theorem 3. *(Serializability) Consider the transition sequence $B \overset{\tilde{\phi}}{\Rightarrow} B'$ with $\mathsf{Act}(B) = \mathsf{Act}(B') = \emptyset$. It is serialized.*

Proof. The proof is by induction on the length n of the transition sequence $B \overset{\tilde{\phi}}{\Rightarrow} B'$. There are two cases.

1. If $n = 1$ then $B \overset{\phi}{\Rightarrow} B'$ and the theorem is vacuously satisfied.
2. If $n > 1$ then $B \overset{\phi_1}{\Rightarrow} B_1 \overset{\phi_2}{\Rightarrow} B_2 ... B_{(n-1)} \overset{\phi_n}{\Rightarrow} B'$. If $f_t(\phi_i) = \emptyset$ for $i = 1, ..., n$ then the condition is vacuously satisfied. Otherwise, let k be the first index for which $f_t(\phi_k) \neq \emptyset$. If $k > 1$ then by inductive hypothesis, the transition sequence $B_{(k-1)} \overset{\tilde{\phi}'}{\Rightarrow} B'$ with $\tilde{\phi}' = \phi_k \phi_{(k+1)} ... \phi_n$ is serialized. Also the transition sequence $B \overset{\tilde{\phi}''}{\Rightarrow} B_{(k-1)}$ with $\tilde{\phi}'' = \phi_1 \phi_2 ... \phi_{(k-1)}$ is serialized. Therefore $B \overset{\tilde{\phi}}{\Rightarrow} B'$ is serialized. Finally, if $k = 1$ then $f_{action}(\phi_1) = t : start$. Let k' be the first index greater than 1 such that $f_{action}(\phi_{k'}) = t : end$. From item (iv) of Prop. 3, $f_{action}(\phi_i) = t : \alpha$ for $i = 2, ...(k' - 1)$ and therefore $B \overset{\tilde{\phi}}{\Rightarrow} B'$ is serialized. Similarly, the transition sequence $\phi_{k'} \phi_{(k'+1)} ... \phi_n$ is serialized. Therefore $B \overset{\tilde{\phi}}{\Rightarrow} B'$ is serialized. $\qquad\square$

Note that the side conditions concerning transactions in the rule redex_t are sufficient conditions to guarantee the properties above. If these conditions are not considered, weaker properties are satisfied. In particular serializability is still valid, but Theorems 1 and 2 are not guaranteed any more.

6   Examples

In this section some examples are reported. They concern:

- *a generic multiple-reactant multiple-product reaction;*
- *the phosphorylation of p53 by a protein kinase;*

– *the citric acid cycle model*, extracted from the KEGG (Kyoto Encyclopedia of Genes and Genomes) database [24].

In the following examples we use the notation x and $\bar{x}$ to indicate respectively the input and the output of an empty message along the channel x.

6.1 Multiple-Reactant Multiple-Product Reactions

Consider a generic reaction R composed of N_r reactants, N_p products and N_m modifiers[5], described by a basal rate r:

$$\kappa_1 A_1 + ... + \kappa_{n_r} A_{n_r} + M_1 + ... + M_{n_m} \xrightarrow{r} \kappa_1' B_1 + ... + \kappa_{n_p}' B_{n_p} + M_1 + ... + M_{n_m}$$

The elements κ_i, κ_j' stand for the stoichiometric coefficients of the reactants and the products, respectively. The total number of reactants is $N_r = \sum_{i=1}^{n_r} \kappa_i$ and the total number of products is $N_p = \sum_{i=1}^{n_p} \kappa_i'$, where n_r and n_p are the number of distinct reactants and products in the reaction, respectively. If we have $N_r + N_m > 2$ or/and $N_p + N_m > 2$ the use of the TBlenX language allows us to overcome the problems we can meet if the standard language is used (see Section 1). The following approach is suggested.

– If there is some further biological information about the reaction, we may decompose it into elementary steps as it happens in reality.
– If further information is not available and elementary steps are unknown, we can translate the reaction in the following way:
  • 1 start event to block the bio-processes involved. The associated rate is the one of the reaction.
  • $(N_r + N_m - 1)$ join events to merge the reactants and $(N_p + N_m - 1)$ split events to get the products;
  • 1 end event to unblock the product-bio-processes.

Globally we need $(N_r + 2N_m + N_p)$ events.

6.2 Phosphorylation of p53 at Serine 15 by Two Enzymes

The second example concerns a simple reaction from Kohn's interaction map [25]. It concerns the phosphorylation of the protein *p53* at a given site, *Ser15* (serine 15), by one protein kinase, either *DNA_pk* or *ATM*. We refer to the translation reported in [14], based on the standard Beta-binders and we show how to represent it into TBlenX. In [14] the phosphorylation is represented by the inter communication between the protein *p53* and one of the two kinases and the following hide of the site representing *Ser15* un-phosphorylated (modeled by a beta binder with subject x) and the unhiding of the site, representing *Ser15* phosphorylated (modeled by a beta binder with subject x'). The elementary biological reaction of phsphorylating a protein, is represented by a sequence of three elementary actions. It would be desirable that these three actions were executed

---

[5] With this term we indicate the species that remain constant in a reaction. For instance, enzymes and inhibitors are modifiers.

atomically to avoid the stop at intermediate steps. For this purpose we translate these reactions into biological transactions. The three proteins may be specified as:

$$B_{p53} = \beta(x, \infty, \{UnPS\,15\})\beta^h(x', \infty, \{PS\,15\})\boldsymbol{B}_1[R|Q_1]$$
$$B_{ATM} = \beta(y, \infty, \Delta_1)\boldsymbol{B}_2[!(\bar{y})^{T_2}|Q_2]$$
$$B_{DNApk} = \beta(z, \infty, \Delta_2)\boldsymbol{B}_3[!(\bar{z})^{T_3}|Q_3]$$

where $R = (x)^{T_1}.(\text{hide}(x), \infty)^{T_1}.(\text{unhide}(x'), \infty)^{T_1}.Q_4$, $T_1 = \{t_{ph1}, t_{ph2}\}$, $T_2 = \{t_{ph1}\}$ and $T_3 = \{t_{ph2}\}$ and $\boldsymbol{B}_i$ for $i = 1, 2, 3$, Q_i for $i = 1, 2, 3, 4$ represent respectively the other beta binders and pi-processes not involved in the reaction. The affinities between the interaction types of the bio-processes representing the *p53* and the two kinases are:

$$\alpha(\{UnPS\,15\}, \Delta_1) = (0, 0, \infty) \quad \alpha(\{UnPS\,15\}, \Delta_2) = (0, 0, \infty)$$

therefore the inter-communications between the protein *p53* and each of the kinase are immediate.

Two start events are defined to specify the start of the two phosphorylations.

$$([1 \diamond B_{p53}, 1 \diamond B_{ATM}] : t_{ph1}; r_{ph1}) \text{ start } ([1 \diamond (B_{p53})^{t_{ph1}}, 1 \diamond (B_{ATM})^{t_{ph1}}])$$

$$([1 \diamond B_{p53}, 1 \diamond B_{DNApk}] : t_{ph2}; r_{ph2}) \text{ start } ([1 \diamond (B_{p53})^{t_{ph2}}, 1 \diamond (B_{DNApk})^{t_{ph2}}])$$

where r_{ph1} and r_{ph2} are the basal rates associated with the two biological interactions. The end events to terminate the transactions are defined as:

$$([1 \diamond (B_P)^{t_{ph1}} : t_{ph1}]; \infty) \text{ end } ([1 \diamond B_P])$$

with $B_P \equiv {}_b\beta^h(x, \infty, \{UnPS\,15\})\beta(x', \infty, \{PS\,15\})$ $\boldsymbol{B}_1[Q_4|Q_1] \parallel B_{ATM}$ and similarly for the second end event, where B_{DNApk} replaces B_{ATM}.

A reduction of the system $S = B_{p53} \parallel B_{ATM} \parallel B_{DNApk} \parallel S'$ is:

$$S \xrightarrow{r_{ph1}, t_{ph1} : start, \epsilon} (B_{p53} \parallel B_{ATM})^{t_{ph1}} \parallel B_{DNApk} \parallel S'$$

$$\xrightarrow{\infty, t_{ph1} : l, \epsilon} (\beta(x, \infty, \{UnPS\,15\})\beta^h(x', \infty, \{PS\,15\})\boldsymbol{B}_1[(\text{hide}(x), \infty)^{T_1}.$$
$$(\text{unhide}(x'), \infty)^{T_1}.Q_4|Q_1] \parallel B_{ATM})^{t_{ph1}} \parallel B_{DNApk} \parallel S'$$

$$\xrightarrow{\infty, t_{ph1} : h, \epsilon} (\beta^h(x, \infty, \{UnPS\,15\})\beta^h(x', \infty, \{PS\,15\})\boldsymbol{B}_1[(\text{hide}(x'), \infty))^{T_1}.Q_4|Q_1]$$
$$\parallel B_{ATM})^{t_{ph1}} \parallel B_{DNApk} \parallel S'$$

$$\xrightarrow{\infty, t_{ph1} : u, \epsilon} (\beta^h(x, \infty, \{UnPS\,15\})\beta(x', \infty, \{PS\,15\})\boldsymbol{B}_1[Q_4|Q_1] \parallel B_{ATM})^{t_{ph1}} \parallel$$
$$B_{DNApk} \parallel S'$$

$$\xrightarrow{\infty, t_{ph1} : end, \epsilon} B_P \parallel B_{DNApk} \parallel S'$$

Note that in TBlenX it is possible to represent the phosphorylation in an alternative way. A unique beta binder can be used to represent the site involved in the phosphorylation (in our example *Ser15*) and the phosphorylation is abstracted by the action *change*.

Initially the type of the site $Ser15$ is $\{UnPS\,15\}$ (i.e. unphosphorylated). The effect of the action is to replace the type $\{UnPS\,15\}$ with the type $\{PS\,15\}$, representing the site phosphorylated. The interaction capabilities of the bio-process are modified.

The new definition of the bio-process representing the protein $p53$ is:

$$B'_{p53} = \beta(x, \infty, \{UnPS\,15\})\boldsymbol{B}_1[(x)^{T_1}.(\mathsf{ch}(x, \{PS\,15\}), \infty)^{T_1}.Q_4|Q_1]$$

The product of the phosphorylation is defined as:

$$B'_P = \beta(x, \infty, \{PS\,15\})\boldsymbol{B}_1[Q_4|Q_1]$$

The start and end events are modified in order to consider these new definitions of the bio-processes representing p53.

A reduction of the system $S_1 = B'_{p53} \parallel B_{ATM} \parallel B_{DNApk} \parallel S'$ is:

$$S_1 \xrightarrow{r_{ph1},t_{ph1}:start,\epsilon} (B'_{p53} \parallel B_{ATM})^{t_{ph1}} \parallel B_{DNApk} \parallel S'$$

$$\xrightarrow{\infty,t_{ph1}:I,\epsilon} (\beta(x, \infty, \{UnPS\,15\})\boldsymbol{B}_1[(\mathsf{ch}(x, \{PS\,15\}), \infty)^{T_1}.Q_4|Q_1] \parallel B_{ATM})^{t_{ph1}} \parallel$$
$$B_{DNApk} \parallel S'$$

$$\xrightarrow{\infty,t_{ph1}:c,\epsilon} (\beta(x, \infty, \{PS\,15\})\boldsymbol{B}_1[Q_4|Q_1] \parallel B_{ATM})^{t_{ph1}} \parallel B_{DNApk} \parallel S'$$

$$\xrightarrow{\infty,t_{ph1}:end,\epsilon} B'_P \parallel B_{DNApk} \parallel S'$$

This approach is more intuitive than the previous one: a unique beta-binder is used to represent the site $Ser15$ of the protein $p53$ and the phosphorylation is abstracted by the change of the type (and therefore the interaction capabilities) of the beta-binder. Note that the action change is peculiar of BlenX and it is not present in the original definition of Beta-binders.

6.3   The Citric Acid Cycle

This model is taken from the *KEGG* metabolic pathway database [24,1]. It regards the *citric acid cycle*, also known as the *Krebs cycle* or *tricarboxylic acid cycle*. This cycle is a fundamental metabolic pathway involving enzymes essential for energy production through aerobic respiration and is also an important source of biosynthetic building blocks used in other processes as for instance the amino acid and the fatty acid biosyntheses.

The biological model. The model consists of a series of chemical reactions of central importance in all living cells that involves a lot of proteins, molecules and enzymes. The citric acid cycle takes place in mitochondria where it oxidizes Acetyl-CoA, derived not only from glycolysis but also from the oxidation of fatty acids. An Acetyl-CoA molecule enters the cycle interacting with Oxaloacetate to create Citrate, for which the subsequent cycle of reactions is named. Acetyl-CoA is oxidized gradually by a chain of reactions. Citrate serves as a substrate for a series of distinct enzyme-catalyzed reactions

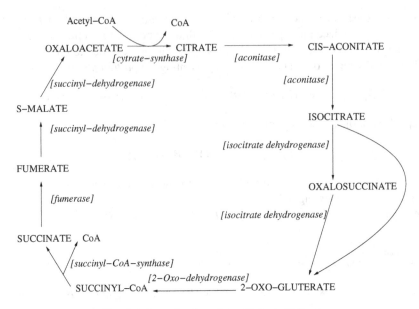

Fig. 3. Schematic view of the Citric Acid Cycle

that occur in sequence and proceed with the formation of intermediate compounds, including Succinate, Fumarate, and S-Malate. S-Malate is converted to Oxaloacetate, which in turn reacts with yet another molecule of Acetyl-CoA, thus producing citric acid and the cycle begins again.

As the cycle proceeds the intermediates are oxidized, transferring their energy to create high-energy electrons in the form of NADH (reduced nicotinamide adenine dinucleotide) or FADH2 (reduced flavin adenine dinucleotide) and one molecule of ribonucleotide GTP (guanosine triphosphate). NADH and FADH2 are coenzymes (molecules that enable or enhance enzymes) that store energy and are passed to a membrane-bound electron-transport chain to produce H_2O. The oxidation of the metabolic intermediates of the pathway also releases two carbon dioxide molecules for each Acetyl-CoA that enters the cycle, leaving the net carbons the same with each turn of the cycle. This carbon dioxide is the one of the sources of CO_2 released into the atmosphere when you breathe.

A schematic representation of the citric acid cycle is reported in Figure 3. Here only the main reactions and species are reported. The species inside the square brackets are the enzymes involved in the reactions, the others are the reactants and products of the reactions in the cycle.

The translation of the model into TBlenX

Initial System. Each *species* is described by a bio-process with one beta binder representing the interaction capabilities of the element. For instance, the element citrate is represented by the following bio-process:

$$B_{Citrate} = \beta(x_{Citrate}, \infty, \{r_{Citrate}\})[\text{nil}]$$

The only pi-process used is nil. Indeed the model gives a high level of abstraction and for representing the reactions of the model we use only join and split events. The other species are translated similarly. The initial system may be given by the bio-processes representing the enzymes, Acetyl-CoA and Oxaloacetate:

$$S = B_{Oxaloacetate} \| B_{Acetyl-Coa} \| B_{CS} \| B_{ID} \| B_A \| B_{SD} \| B_{2HD} \| B_{SCS} \| B_F$$

The enzymes are denoted by the initial of the names in capital letters. So, for instance, the bio-process B_{CS} represents the enzyme *Citrate Synthase*. For simplicity we consider only an element for each species.

Reactions. Each *reaction* is rendered by a set of suitable join and split events. We consider here biological transactions to represent each reaction atomically. The translation of the main reactions follows.

– The first kind of reaction is the enzymatic reaction with one reactant, one modifier and one product. It is translated by using a start event to block the bioprocess involved, followed by a join event and a split event. Finally, an end event unblocks the bio-processes returning the final products. All the reactions with the exception of two cases are of this kind. We show how to translate one, the others are dealt with in the same way. We consider the reaction:

$$Fumerate + Fumerase \xrightarrow{r_1} S\text{-}Malate + Fumerase$$

where r_1 is the basal rate associated with the reaction. Fumarate undergoes a hydration catalyzed by Fumarase to produce S-Malate. Let $B_{Fumerate}$, B_F and $B_{S-Malate}$ be the bio-processes representing Fumerate, Fumerase and S-Malate, respectively. The start of the reaction blocks the element $B_{Fumerate}$ and B_F and it is described by the event:

$$([1 \diamond B_{Fumerate}, 1 \diamond B_F] : t_1; r_1) \text{ start } ([1 \diamond (B_{Fumerate})^{t_1}, 1 \diamond (B_F)^{t_1}])$$

The name t_1 indicates the transaction. One join event is used to represent the formation of the intermediate element $complex(Fumerate, F)$ described by the bio-process $B_{c(Fumerate,F)}$:

$$([1 \diamond (B_{Fumerate})^{t_1}, 1 \diamond (B_F)^{t_1}]; \infty) \text{ join } ([1 \diamond (B_{c(Fumerate,F)})^{\{t_1\}}])$$

A split is used to get the products and it is described by:

$$([1 \diamond (B_{c(Fumerate,F)})^{\{t_1\}}]; \infty) \text{ split } ([1 \diamond (B_{S-Malate})^{\{t_1\}}, 1 \diamond (B_F)^{\{t_1\}}])$$

Finally, the event end to unblock the elements is:

$$([1 \diamond (B_{S-Malate})^{\{t_1\}}, 1 \diamond (B_F)^{\{t_1\}}] : t_1; \infty) \text{ end } ([1 \diamond B_{S-Malate}, 1 \diamond B_F]).$$

– The second kind of reaction is described by two reactants, two products and one modifier. In the model there is only one reaction of this kind and it is:

$$acetyl\text{-}CoA + Oxoloacitate + CS \xrightarrow{r_2} CoA + Citrate + CS$$

where *CS* stands for for *Citrate Synthase*. *Acetyl-CoA* interacts with *Oxaloac-etate* to form *Citrate* and *CoA*. The reaction is translated by using one start to block the bio-processes representing the reactants and the modifier, two join events to form the intermediate complexes, two split events to get the products and one end to unblock the elements.

– The last kind of reaction is represented by:

$$Oxolossucinate + ID \xrightarrow{r_3} 2\text{-}Oxo\text{-}glutarate + CO_2 + ID$$

where *ID* stands for *Isocitrate Dehydrogenase*. This enzyme catalyzes the re-action from *Oxalosuccinate* to *CO2* and *2-Oxo-glutarate*. In this case it is nec-essary to define a start event, followed by a join and two split events and finally one end event.

Alternative approach. The translation of a biochemical network at a high level of abstraction by using join and split events is intuitive, but it is not always the more convenient approach. A major drawback is that a lot of events must be defined. In the case of the citric acid cycle, we globally need 47 events: 11 start, 11 end, 12 join and 13 split events. An alternative way to represent biochemical reactions is to consider the actions *complex* and *decomplex* instead of the events *join* and *split*. This possibility has been introduced in the BlenX language. A discussion about the advantages of the use of complex and decomplex actions is reported in [39].

In the following we report the main ideas about the translation of the citric acid cycle by using this approach. Each biological element is represented by a bio-process. Each bio-process is characterized by a beta binder, whose type indicates the interaction capabilities of the element and, differently from before, pi-processes instead of events describe the interactions between biological species. The events considered are only the ones describing the start and the end of the transactions and one split for describing the last kind of reaction.

We describe briefly how to translate each reaction of the citric acid cycle. The trans-lation of the first kind of reactions is based on the following idea: the reactant forms a complex with the enzyme and after that the reactant changes its interaction capabilities (i.e. the type associated with the interaction site) and is transformed into the product. Then we have the decomplexation of the product from the enzyme. A biological trans-action is used to make the sequence atomic.

Consider in detail the reaction $Fumerate + Fumerase \xrightarrow{r_1} S\text{-}Malate + Fumerase$. We can specify the reactant (Fumerate) and the enzyme (Fumerase) in the following way:

$$B'_{Fumerate} = \beta(x, \infty, \Gamma_{Fumerate})[(\bar{x})^{\{t_1\}}.(\mathsf{ch}(x, \Gamma_{\text{S-Malate}}), \infty)^{\{t_1\}}.P_{\text{S-Malate}}]$$
$$B'_F = \beta(x_F, \infty, \Gamma_F)[P_F]$$

where $P_{\text{S-Malate}}$ is the pi-process associated with the product S-Malate and $P_F = P_{F1}|!(f)^{\{t_1\}}.P_{F1}$ with $P_{F1} = (x_F.\bar{f}.\mathsf{nil})^{\{t_1\}}$. We must define the affinities between the interaction types in an appropriate way. Specifically we have:

$$\alpha(\Gamma_{Fumerate}, \Gamma_F) = (\infty, 0, \infty) \quad \alpha(\Gamma_F, \Gamma_{\text{S-Malate}}) = (0, \infty, 0)$$

In the first case the rates associated with the complexation and the inter-communication are immediate and decomplexation is not possible, while in the latter case we have that only decomplexation is possible and it is immediate. The start and the end of the transaction t_1 are defined as in the previous approach, but here we consider the new definitions for the elements involved in the reaction. The reduction of the system $S = B'_{Fumerate} \parallel B'_F$ is the following:

$$S \xrightarrow{r_1, t_1 : start, \epsilon} (B'_{Fumerate} \parallel B'_F)^{t_1}$$

$$\xrightarrow{\infty, t_1 : C, \epsilon} (\beta(x, \infty, \Gamma_{Fumerate})^c[(\bar{x})^{\{t_1\}}.(\mathsf{ch}(x, \Gamma_{\text{S-Malate}}), \infty)^{\{t_1\}}.P_{\text{S-Malate}}] \parallel$$
$$\beta(x_F, \infty, \Gamma_F)^c[P_F])^{t_1}$$

$$\xrightarrow{\infty, t_1 : I_c, \epsilon} (\beta(x, \infty, \Gamma_{Fumerate})^c[(\mathsf{ch}(x, \Gamma_{\text{S-Malate}}), \infty)^{\{t_1\}}.P_{\text{S-Malate}}] \parallel$$
$$\beta(x_F, \infty, \Gamma_F)^c[(\bar{f}.\mathsf{nil})^{\{t_1\}}|!(f)^{\{t_1\}}.P_{F1}])^{t_1}$$

$$\xrightarrow{\infty, t_1 : C, \epsilon} (\beta(x, \infty, \Gamma_{\text{S-Malate}})^c[P_{\text{S-Malate}}] \parallel \beta(x_F, \infty, \Gamma_F)^c[(\bar{f}.\mathsf{nil})^{\{t_1\}}|!(f)^{\{t_1\}}.P_{F1}])^{t_1}$$

$$\xrightarrow{\infty, t_1 : Dc, \epsilon} (\beta(x, \infty, \Gamma_{\text{S-Malate}})[P_{\text{S-Malate}}] \parallel \beta(x_F, \infty, \Gamma_F)[(\bar{f}.\mathsf{nil})^{\{t_1\}}|!(f)^{\{t_1\}}.P_{F1}])^{t_1}$$

$$\xrightarrow{\infty, t_1 : i, \epsilon} (\beta(x, \infty, \Gamma_{\text{S-Malate}})[P_{\text{S-Malate}}] \parallel \beta(x_F, \infty, \Gamma_F)[!(f)^{\{t_1\}}.P_{F1}|P_{F1}])^{t_1}$$

$$\xrightarrow{\infty, t_1 : end, \epsilon} B'_{\text{S-Malate}} \parallel B'_F$$

where $B'_{\text{S-Malate}} = (\beta(x, \infty, \Gamma_{\text{S-Malate}})[P_{\text{S-Malate}}]$ is the bio-process representing the S-Malate.

With regards to the second kind of reaction, we follow a similar approach: we have a first complex action between the two reactants (Acetyl-CoA and Oxoloacitate) and then another complex action between one reactant and the enzyme (CS). After complexation, the two reactants change the interaction capabilities and become products (CoA and Citrate). A start event is used to block the elements involved in the reaction and one end is used to unblock them after the execution of the intermediate steps.

The last kind of reaction is translated similarly. We have one complex action between the reactant and the enzyme followed by a decomplexation to obtain the enzyme and the complex of the two products. Finally, a split event is necessary to get the two products.

7 Discussion and Conclusion

The BlenX language is a formalism recently defined for the modelling and analysis of biological systems. One drawback of this language is the modelling of multiple-reactant multiple-product reactions. Indeed, in order to describe this kind of reactions into this language we should decompose them into binary elementary reactions. Generally, there are different ways to decompose a reaction and it is not possible to say what is the most appropriate solution. Furthermore, by considering this approach the reaction could stop at intermediate steps leading to a deadlock. This may happen for instance when a reactant involved in the reaction is missing. Last but not least, the dynamics of the reaction is given in terms of a global rate and, therefore, there is the problem of how to find the rates for the elementary steps in which the reaction may be decomposed.

The aim of this work is to extend **BlenX** in order to model a sequence of actions representing complex reactions as if it were atomic and therefore to represent complex reactions with more than two reactants or more than two products in a suitable way with binary interactions. At this purpose we extended **BlenX** with biological transactions. The extension with biological transactions presented in this paper matches these specifications, as explained in the following items.

- In this extension reactions are decomposed into elementary steps. However in this case the order is not important, as the interactions involving the elementary reactions are internal to the transactions. The relevant actions are the start and the end of the transactions, the other ones are only auxiliary.
- From the definition of biological transactions given in this paper, it follows that when the transaction starts it completes. Therefore it is not possible that a reaction stops at intermediate steps.
- The global rate of the reaction is associated to the start action. In this way the actual rate of the reactions depends on all the reactants involved.The other actions follow as immediate and it is not necessary to find a rate for each elementary step.

From the observations above it is clear that **TBlenX** is useful to deal with complex reactions whose details are unknown.

Some final observations concern the use of Gillespie's algorithm and the kind of rate. Firstly, in this work we used the Gillespie's method as the reference stochastic algorithm. We referred to the extended version that considers reactions with more than two reactants, widely used in the simulation tools. Obviously, other stochastic algorithms could be considered as well.

Secondly, we assumed for simplicity that the kinetics associated with a reaction is always expressed through the mass-action law and we can associated each action with a constant basal rate. However, when complex reactions are considered the dynamics of the reaction may be described by more complex laws than the well-know mass-action. The application of the Gillespie's algorithm to these complex kinetic laws has been recently discussed and formalized [36,10]. These approaches are approximations and are based on some assumptions. The **BlenX** language supports the definition of general kinetic laws in terms of rate functions (see [19]). These are used to calculate the actual rate of the associated reaction. The addition of general kinetic laws to our extension is straightforward and it is similar to the case of mass-action proposed here. Specifically we can associate the rate function with the start action and consider it in the derivation of the actual rate.

Finally, we recall that this transaction mechanism is under implementation in the Beta Workbench [2].

Acknowledgements

This work was done while the author was working towards her PhD degree at the University of Trento (Italy). The author would like to thank Corrado Priami for his helpful comments.

References

1. KEGG home page, http://www.genome.jp/kegg
2. The Beta Workbench, http://www.cosbi.eu/Rpty_Soft_BetaWB.php
3. Bocchi, L., Laneve, C., Zavattaro, G.: A calculus for long-running transactions. In: Najm, E., Nestmann, U., Stevens, P. (eds.) FMOODS 2003. LNCS, vol. 2884, pp. 124–138. Springer, Heidelberg (2003)
4. Bortolussi, L., Policriti, A.: Modeling Biological Systems in Stochastic Concurrent Constraint Programming. Constraints 13 (2008)
5. Bruni, R., Laneve, C., Montanari, U.: Orchestrating transactions in join calculus. In: Brim, L., Jančar, P., Křetínský, M., Kucera, A. (eds.) CONCUR 2002. LNCS, vol. 2421. Springer, Heidelberg (2002)
6. Busi, N., Zavattaro, G.: On the Serializability of Transactions in JavaSpaces. In: Proc. of CONCOORD 2001. LNCS, vol. 54. Springer, Heidelberg (2001)
7. Butler, M., Ferreira, C.: An operational semantics for StAC, a language for modelling long-time business transactions. In: De Nicola, R., Ferrari, G.L., Meredith, G. (eds.) COORDINATION 2004. LNCS, vol. 2949. Springer, Heidelberg (2004)
8. Butler, M., Hoare, T., Ferreira, C.: A trace semantics for long-running transactions. In: Abdallah, A.E., Jones, C.B., Sanders, J.W. (eds.) Communicating Sequential Processes. LNCS, vol. 3525, pp. 133–150. Springer, Heidelberg (2005)
9. Calder, M., Gilmore, S., Hillston, J.: Modelling the influence of RKIP on the ERK signalling pathway using the stochastic process algebra PEPA. T. Comp. Sys. Biology VI, 1–23 (2006)
10. Cao, Y., Gillespie, D., Petzold, L.: Accelerated Stochastic Simulation of the Stiff Enzyme-Substrate Reaction. J. Chem. Phys. 123, 144917–144929 (2005)
11. Ciocchetta, F., Hillston, J.: Bio-PEPA: a framework for the modelling and analysis of biological systems, Tech. Rep. EDI-INF-RR-1231, School of Informatics, University of Edinburgh (2008)
12. Ciocchetta, F., Hillston, J.: Bio-PEPA: an extension of the process algebra PEPA for biochemical networks. In: Proc. of FBTC 2007. Electronic Notes in Theoretical Computer Science, vol. 194 (2008)
13. Ciocchetta, F., Priami, C.: Biological transactions for quantitative models. In: Proc. of MeCBIC 2006. ENTCS, vol. 171 (2007)
14. Ciocchetta, F., Priami, C., Quaglia, Q.: Modeling Kohn Interaction Maps with Beta-Binders: An Example. T. Comp. Sys. Biology III, 33–48 (2005)
15. Danos, V., Krivine, J.: Formal molecular biology done in CCS-R. In: BioConcur 2003, Workshop on Concurrent Models in MolecularBiology (2003)
16. Degano, P., Priami, C.: Enhanced operational semantics: A tool for describing and analysing concurrent systems. ACM Computing Surveys 33(2), 135–176 (2001)
17. Dematte', L., Priami, C., Romanel, A.: BetaWB: modelling and simulating biological processes. In: Proceedings of Summer Computer Simulation Conference (SCSC 2007) (2007)
18. Dematte', L., Romanel, A., Priami, C.: The beta workbench: a computational tool to study the dynamics of biological systems. Briefings in Bioinformatics (to appear)
19. Dematte', L., Romanel, A., Priami, C.: The blenx language: A tutorial. In: Bernardo, M., Degano, P., Zavattaro, G. (eds.) SFM 2008. LNCS, vol. 5016, pp. 313–365. Springer, Heidelberg (2008)
20. Fages, F., Soliman, S., Chabrier-Rivier, N.: Modelling and querying interaction networks in the biochemical abstract machine BIOCHAM. Journal of Biological Physics and Chemistry 4(2), 64–73 (2004)
21. Gillespie, D.: Exact stochastic simulation of coupled chemical reactions. J. Phys. Chem. 81(25), 2340–2361 (1977)

22. Hillston, J.: A Compositional Aprroach to Performance Modelling. Cambridge University Press, Cambridge (1996)
23. Hucka, M., Finney, A., Bornstein, B., Keating, S., Shapiro, B., Matthews, J., Kovitz, B., Schilstra, M., Funahashi, A., Doyle, J., Kitano, H.: Evolving a Lingua Franca and Associated Software Infrastructure for Computational Systems Biology: The Systems Biology Markup Language (SBML) Project. Systems Biology 1(1) (2004)
24. Kanehisa, M., Goto, S.: KEGG: Kyoto encyclopedia of genes and genomes. Nucleic Acids Research 28(1), 27–30 (2000)
25. Kohn, K.: Molecular Interaction Map of the Mammalian Cell Cycle Control and DNA repair Systems. Molecular Biology of the Cell 10, 2703–2734 (1999)
26. Kuttler, C., Niehren, J., Blossey, R.: Gene regulation in the π-calculus: simulating cooperativity at the lambda switch. In: Second international workshop on concurrent models in molecular biology (BioConcur 2004) (2004)
27. Laneve, C., Zavattaro, G.: Foundations of web transactions. In: Sassone, V. (ed.) FOSSACS 2005. LNCS, vol. 3441, pp. 282–298. Springer, Heidelberg (2005)
28. Laneve, C., Zavattaro, G.: Web-π at work. In: De Nicola, R., Sangiorgi, D. (eds.) TGC 2005. LNCS, vol. 3705. Springer, Heidelberg (2005)
29. Lecca, P., Priami, C.: Cell Cycle control in Eukaryotes: a BioSpi model. In: Bioconcur 2003. ENTCS, vol. 180 (2007)
30. Lecca, P., Priami, C., Quaglia, P., Rossi, B., Laudanna, C., Costantin, G.: Language modeling and simulation of autoreactive lymphocytes recruitment in inflamed brain vessels. SIMULATION: Transactions of The Society for Modeling and Simulation International 80(4), 273–288 (2004)
31. Nagasaki, M., Onami, S., Miyano, S., Kitano, H.: Bio-calculus: Its concept and molecular interaction. In: Asai, K., Miyano, S., Takagi, T. (eds.) Genome Informatics 1999, vol. 10. Universal Academy Press (1999)
32. Prandi, D.: A formal study of biological interactions, Ph.D. thesis, university of Trento (2006)
33. Priami, C., Quaglia, P.: Beta binders for biological interactions. In: Danos, V., Schachter, V. (eds.) CMSB 2004. LNCS (LNBI), vol. 3082, pp. 20–33. Springer, Heidelberg (2005)
34. Priami, C., Regev, A., Silverman, W., Shapiro, E.: Application of a stochastic name-passing calculus to representation and simulation of molecular processes. Information Processing Letters 80(1), 25–31 (2001)
35. Priami, C., Romanel, A.: The Decidability of the Structural Congruence for Beta-binders. In: Proc. of MeCBIC 2006. ENTCS, vol. 171, pp. 155–170 (2007)
36. Rao, C., Arkin, A.: Stochastic chemical kinetics and the quasi-steady-state assumption: application to the Gillespie algorithm. Journal of Chemical Physics 11(11) (2003)
37. Regev, A., Panina, E.M., Silverman, W., Cardelli, L., Shapiro, E.: BioAmbients: An Abstraction for Biological Compartments. Theoretical Computer Science 325(1), 141–167 (2004)
38. Regev, A., Silverman, W., Shapiro, E.: Representation and simulation of biochemical processes using the π-calculus process algebra. In: Proceedings of the Pacific Symposium of Biocomputing 2001, vol. 6 (2001)
39. Romanel, A., Dematte', L., Priami, C.: The Beta Workbench, Tech. Rep. TR-03-2007, The Microsoft Research-University of Trento Centre for Computational and Systems Biology (2007)
40. Wolkenhauer, O., Ullah, M., Kolch, W., Cho, K.H.: Modelling and Simulation of Intra-Cellular Dynamics: Choosing an Appropriate Framework. IEEE Transactions on NanoBioScience 3(3), 200–207 (2004)

Author Index